DESIGN THINKING

DESIGN THINKING
THE HANDBOOK

Falk Uebernickel
Li Jiang
Walter Brenner
Britta Pukall
Therese Naef
Bernhard Schindlholzer

Published by

WS Professional, an imprint of
World Scientific Publishing Co. Pte. Ltd.
5 Toh Tuck Link, Singapore 596224
USA office: 27 Warren Street, Suite 401-402, Hackensack, NJ 07601
UK office: 57 Shelton Street, Covent Garden, London WC2H 9HE

Library of Congress Cataloging-in-Publication Data
Names: Uebernickel, Falk, author. | Jiang, Li, author. | Brenner, Walter, author.
Title: Design thinking : the handbook / Falk Uebernickel (University of St. Gallen, Switzerland), Li Jiang (Stanford University, USA), Walter Brenner (University of St. Gallen, Switzerland), Britta Pukall (Milani Design & Consulting, Switzerland), Therese Naef (Milani Design & Consulting, Switzerland), Bernhard Schindlholzer (Google, Switzerland).
Description: 1 Edition. | New Jersey : World Scientific, [2019] | Translation from German: Design Thinking : Das Handbuch. | Includes bibliographical references and index.
Identifiers: LCCN 2019005287 | ISBN 9789811202148 | ISBN 9789811203503 (pbk)
Subjects: LCSH: Project management. | New products. | Management—Technological innovations.
Classification: LCC HD69.P75 U4313 2019 | DDC 658.4/063—dc23
LC record available at https://lccn.loc.gov/2019005287

British Library Cataloguing-in-Publication Data
A catalogue record for this book is available from the British Library.

ISBN 978-9-811-20214-8
Book and cover design: milani design & consulting AG, Thalwil, Schweiz
Type setting and lithography: Red Cape Production, Berlin
Translation: Rachel Ames-Brooks, Lawrence McGrath
Proofreading and addional editing: Padraic Convery
Project management: Andreas Vogel

Desk Editor (WSPC): Shreya Gopi

Printed in Singapore

The English version of the book would have never been possible without the help of these awesome people: Rachel Ames-Brooks, Lawrence McGrath, Karina Fassbender, Laura Ferreira, Andreas Vogel, and Janine Milstrey. Thank you all for your passion and dedication!

THIS BOOK

This book is for those who want to generate ideas and create innovation in their organizations in the fastest and most practical way possible. In a cookbook manner, this book provides recipes that can be implemented immediately in daily business.

Design thinking is an innovation method that focuses on the customer in order to generate product, service and business model innovations beyond the well-known paths.

The authors themselves know design thinking not only from the academic, scientific world, but from numerous long-term applications at companies in a range of industries.

CONTENTS

SETUP

INTRODUCTION AND READING INSTRUCTIONS

Families and couples often live for many decades in the same apartment or house. How can you design furniture that is attractive to both young people and the elderly, including those with disabilities?

Errors in the hospital system can be life-threatening for patients. How can you significantly improve the traceability of samples through digitization?

You have developed a drug for people in emerging markets. It must not cost more than 1 euro per bottle. The available components already cost more than 2 euros today. How do you solve this problem?

In coming years, high-value assets will be bequeathed by parents to their descendants in industrialized countries. Often, however, financial institutions have little attachment to the younger generation. How can that be addressed?

We have been allowed to deal intensively with these and many other questions over the past few years. At first glance, they seem to have no connecting element. From the finance and insurance sector to pharmaceutical companies, but also from a functional point of view, these questions cover a broad spectrum of topics, from marketing through product management to research and development. The focus is on product, service and business model innovations. All these questions have in common that the companies and organizations behind them consider fresh, new perspectives on these topics important and place customers and other parties — the so-called stakeholders — at the center of finding solutions. It requires empathy to consider problems and challenges from the customer's point of view, which in turn creates new possibilities for ideas and solutions. The process and the way of thinking by which these questions are dealt with is called design thinking.

This handbook is a practical guide for beginners and people who have already gained experience in design thinking and would like to get to know new perspectives on the methodology. And for students, it provides a great practical basis to learn the skills and methods involved.

Chapter 1 provides a brief but comprehensive overview of all the elements and tools essential to design thinking. Chapter 2, the toolbox, focuses on individual tools that are often used in design thinking projects. Chapter 3 gives an insight into our understanding of design thinking spaces and how they should be structured. Finally, chapters 4 and 5 present practical experiences of introducing design thinking in companies using a transformation model and real-world case studies.

The colored markings on the margins should help you to quickly look up and locate the methods you are looking for.

THANK YOU!

This book would never have come about without the help of many people and the support of several companies. This list is certainly not exhaustive, but our special thanks go to the following people:

Jagat Adhiya, Manuel Ailinger, Samuel Beer, Henrik Beckmann, Kristen Bennie, Katharina Berger, Fernando Bernal, Marc Binder, Sophie Bürgin, Daniel Cantieni, Tamara Carleton, William Cockayne, Alexander Eck, Martin Eppler, Christopher Este, Niels Feldmann, Michelle Gaegauf, Nina Gandt, Stefan Gaus, Alexander Grots, Ramona Günzel, Marie-Christine Jaeger Firmenich, Jennifer Hehn, Jennifer Heier, Joscha Held, Friederike Hoffmann, Manuel Holler, Oliver Kempkens, Sebastian Kernbach, Marc Kohler, Michael Larsson, Volker Laska, Larry Leifer, Hans Tobias Macholdt, Bettina Maisch, Christian Michels, Boris Milkowski, Janine Milstrey, Susann Müller, Timm Püller, Axel Raidt, Barbara Rohner, Gerhard Satzger, Eric Schmid, Moritz Schnewlin, Gerhard Schwabe, Frank Seifert, Philipp Skribanowitz, Christian Steiger, Henning Strobel, Marion Uebernickel, Christophe Vetterli, Matteo Vignoli, Livia Weder, Andrea Weierich

We would also like to thank the design thinking alumni and students of the Universities of St. Gallen and Zurich. Without you, we would not have achieved such fantastic results in recent years:

Tiziana Aiolfi, Dominik Alvermann, Amanda Bachmann, Nicolas Beck, Nicolas Berchten, Nina Birri, Fabian Bischof, Jasmin Bissig, Tobias Blum, Maximilian Bredow, Andreas Jürg Breitenmoser, Ivo Brennwald, Marco Brunori, Michel Bürki, Andrei Cojocariu, Carlo De Pascalis, Michael Denzler, Caroline Dohle, Franziska Dolak, Philipp Elbel, Jessica Enstedt, Marco Eugster, Philipp Fleckner, Tobias Fuhrimann, Laura Galamb, Marie-Charlotte Gasser, David Geisser, Simon Gensmer, Tobias Giger, Matthias Gisler, Rouven Emanuel Gruenig, Naomi Haefner, Christian Haeuter, Tatiana Hanz, Joscha Held, Nadine Hergovits, Anja Hovorka, Samuel Huber, Matthias Hübner, Wanja Humanes, Beatrice Hutzli, Tobias Imwinkelried, Dominik Jocham, Alexander Jung-Loddenkemper, Pascal Kappeler, Astrit Kazimi, Patrick Keil, Will Kölbener, Yair Kollmann, Nicolas Kuhn, Daniel Alexander Kunz, Florian Kunz, Joel Kurmann, Adrienne Kristin Lock, Mario Malzacher, Camille Martinache, Tiziana Mauchle, Milena Mend, Petra Monn, Alexander Muelli, Franziska Müller, Arjun Muralidharan, Edouard Papaux, Stephanie Petersen, Vanessa Pinter, Christine Popp, Carolyn Ragaz, Martina Rakaric, Marc Rieben, Rico Rinderknecht, Manuela Risch, Deborah Schaub, Markus Schillinger, Dominic Schlegel, Marc Schlegel, Manuel Schöni, Florim Shabani, Martin Spielmann, Patricia Steffen, Maximilian Steinbach, Andreas Stockburger, Emanuel Stoeckli, Daniel Strebel, Simon Streit, Arlene Struijk, Raphael Stücheli, Carina Them, Raphael Thommen, Lukas Troxler, Michelle Martina Tschumi, Timo Van Bargen, Eliane Vancura, Andreas Vogel, Pasqual Vossberg, Harald Weishuber, Johannes Weiss, Dominic Widmer, Marianne Elisabeth Wiesli, Melissa Willhaus, Matthias Wittmann, Nikola Zic, Dominik Zurbuchen, Martin Zwahlen

1

METHOD OVERVIEW

What constitutes design thinking and what does such a project entail? An overview of the project cycle and the tools it uses.

WHAT IS DESIGN THINKING?

Design thinking is an innovation method that uses an iterative process to deliver user- and customer-oriented results to solve complex problems. The term “design” has Anglo-Saxon roots. In contemporary German-speaking regions, the term “design” is mainly associated with the artistic, creative and especially formative and compositional activities in workplaces. However, in today’s English-speaking areas the term also covers the conceptual and technical creation of systems and objects. As a method, design thinking thus relates to a far broader field than the purely aesthetic components of design. It is a method for engineers by engineers.

Design thinking opens a door for people of all age groups, who are (once more) playful and curious about looking at and working on problem statements. The method lets us think about the apparently illogical and unachievable, seriously discuss them, and arrive at surprising realizations. Unfortunately, this ability is often removed by our upbringing and educational institutions (Creuznacher & Grots, 2011).

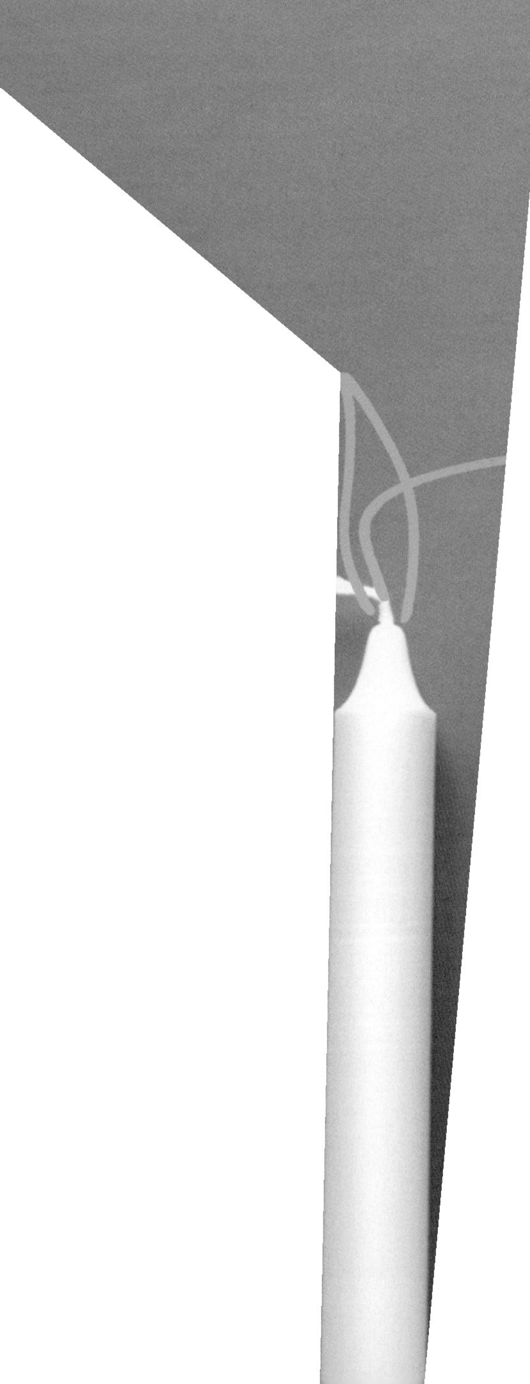

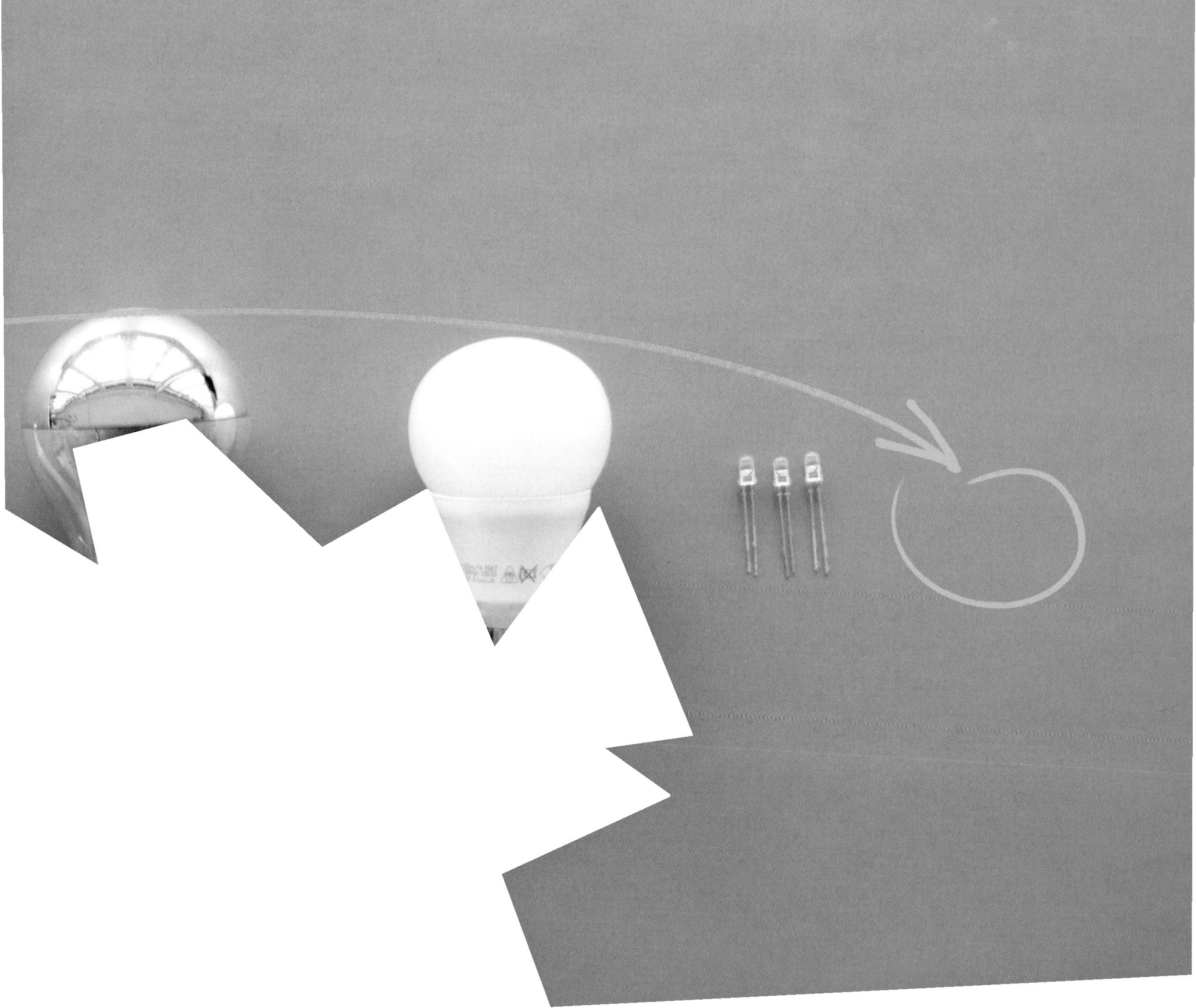

WHAT IS DESIGN THINKING?

The design thinking method is integrated into a cultural transformation of society — what Richard Florida calls "The Rise of the Creative Class" (Florida, 2014). Together with the technological advances of the western world, this has had a substantial impact. This shift reflects a desire and demand for creativity. Not only in society, but also in the workplace, creativity has become an economic requirement. Subjective desires and social expectations of creativity thus come together — "one wants to be and should be creative" (*"man will kreativ sein und soll es sein"*) (Reckwitz, 2012). Design thinking stands for making the new and is thus the embodiment of creativity. The method fits current times because it helps people make creativity manageable.

Basic principles

Design thinking is based on principles that are lived out in every project and department in which they are employed, independent of process implementation:

Empathy: It is decisive in design thinking that people summon empathy for other people so they can better understand problems and develop better solutions. Empathy is not only central to interactions with clients, but also to internal communication in teams. It helps us to understand a dialogue from the point of view of our conversation partner, and supports us in interpretation (Meinel, Plattner & Weinberg, 2009).

Fail forward: Failure is a fundamental part of the innovation process! Unfortunately, the word "failure" is generally used negatively to express breakdowns and a person's lack of ability. Starting early, most of us were raised in a no-failure culture. Organizations have established countless assessment criteria. In contrast, design thinking sees failure and missteps as central parts of a learning process. Through recognition and evaluation of their failures, design thinking teams can correct them more quickly during projects and thus reduce risks.

Fail often and therefore early: Connected with the principle of "fail forward", the design thinking method provokes failures early in the course of the project through short iterative cycles—particularly those failures that are most common. If failure is seen as a source of knowledge, failing often and early leads also to an accelerated learning process within the team (see Leifer & Steinert, 2011).

Autonomy: Recent creativity research clearly shows that the people's subjective perception of freedom of action plays a definitive role in deciding the success of a project (see Amabile, Conti, Coon, Lazenby & Herron, 1996). Frustration and demotivation are primarily found among team members in restrictive environments, where approaches to problem-solving, thinking or even complete solutions are determined in advance. In such an environment, creativity diminishes significantly. Therefore, a basic principle of the design thinking method is that teams have the highest possible autonomy in decision-making and action-taking.

Test with your customer and user: Customers and users validate prototypes—not management or the project director! Operationally and methodologically, this means that prototypes of ideas are tested early with clients and users in a design thinking project. With the help of the results, teams can make decisions about the continued development of the idea, or rejection. This principle also actively fosters understanding of the client (Leifer & Steinert, 2011).

Collecting feedback: "Yes, but..." is the standard phrase in many organisation's workshops and meetings. In conversations, this phrase acts as a brake on the flow of ideas. Ultimately, the use of the phrase shows only that one would rather express doubt and concern instead of going into the idea with his or her conversation partner. However, design thinking teams try to give elaborative, and thus connected, constructive feedback. An effective technique is to say, "Yes, and...". This technique leads one to engage with his or her conversation partner, and give feedback that helps to further develop the idea.

Make it tangible: In design thinking, all results should be visible and tangible in the form of a prototype. While ideas do not exist in reality, prototypes enable them to be seen and experienced (Brown, 2009). While demand for tangibility in technical systems is understandable, when it comes to services, this demand may appear absurd. Scores of projects with service organizations have nonetheless shown us that services can also be made tangible and comprehensible—by storytelling, for example. Furthermore, it has been shown that in many situations, prototyping contributes to making the complexity of a problem manageable for design thinking teams (Uebernickel & Brenner, 2015).

Interdisciplinarity: Design thinking lives off different perspectives and interpretations. Teams are therefore put together with attention to interdisciplinarity. Often, team members have different educational backgrounds—for example, engineering, architecture, design, business administration, etc. (Meinel, Plattner & Weinberg, 2009). The team can benefit from expertise in various subjects, but also the different strategies of its individual members.

Optimistic and curious: Team members must have the internal impetus to search for new question statements and approaches to problem-solving. Even when the project's procedure appears to be set, team members should not lose their will to develop a good solution.

Experimental: Design thinking is a pragmatic approach that prefers experiment to theory. Concretely, this means that in projects, solutions are developed and tried out early with the help of prototypes, and the client and tester decide on a given prototype's success.

WHAT IS DESIGN THINKING?

The history of design thinking

The design thinking method introduced in this book was developed in the '70s and '80s at Stanford University in Palo Alto, California. In the context of educating engineers, professors noticed that focusing purely on the technological aspects of their students' training was not enough to prepare them for the market needs and challenges of the future. While traditional training focused on the "what" and "how", explanations of "what for" and "why" were missing. However, these exact two questions are essential in order not only to develop technically challenging innovations, but also to successfully place them on the market. The recognition of purpose, that is, for what an innovation is necessary, marks the current success of the design thinking method. Based on this recognition, the curriculum for engineers at Stanford University was extended two decades ago, placing the customer, as the product and service user, at the center.

Subsequently, a whole design thinking method set was developed, drawing strongly from the root ideas of the earlier foundational period. This set of methods considered human psychological factors (desirability), as well as technical and processual factors (feasibility), and economic factors (viability) (Brown, 2009). Design thinking balances all three dimensions, but places the human psychological factor at the center—following the motto "innovations are by people, for people". That said, economic and technological factors are not neglected, rather their content is aligned around the human component.

The question remains—for which kinds of problems is design thinking appropriate? In the scientific community, we distinguish between at least two classes of problems: "simple or clearly structured problems" (tame problems) and "complex or multi-layered problems" (wicked problems) (Buchanan, 1992). Both classes of problems can be successfully addressed using design thinking.

Clear problems (tame problems)

Tame problems and problem statements are characterized by an ease of definition and specification. The starting condition (A) and goal condition (B) are both comprehensible. Accordingly, most people agree on the situation in tame problems, and root cause analysis is comparatively easy. However, tame problems still often require creative solutions.

In tame problems, design thinking focuses more on the analysis of customer needs and the generation of ideas and prototypes than deep background analysis. It is not the final goal to develop disruptive solutions, but rather to implement incremental yet creative developments.
In practice, design thinking has shown itself especially to impact speed, agility and client orientation in projects. With this approach, we developed new service center processes and software solutions for a German bank. In this

project we profited from the iterative approach that design thinking enabled. The short cycles connected with frequent client interaction resulted in an innovative solution in a short time. The design thinking process served to limit drawn-out discussions about abstract concepts. Instead, we quickly made prototypes and placed them before service center employees in decision-making roles.

Complex, multi-layered and diffuse problems (wicked problems)

In contrast to tame problems, wicked problems are difficult to name and define. The client encounters difficulties in explaining the problem and determining its cause. Depending on how the problem is grasped and understood, different potential solutions could lead in completely different directions. Therefore, there is generally no consensus between people regarding what is causing the problem, even when they work within the same organization. Another characteristic of this problem class is that possible solutions cannot be classified as right or wrong. At best, they can be placed on a scale from good to bad. Additionally, wicked problems are frequently dependent on, and of mutual influence with, other wicked problems (Rittel & Webber, 1973). The question of how heavily-indebted people can be supported in becoming debt-free is exemplary. Answering this question is very complex and has posed a challenge to experts for decades. There will not be just a single solution to this problem.

From the beginning, wicked problems require an approach such as that which design thinking enables. Heeding all possible perspectives, the problem space must be explored and defined before drawing any single solution into consideration at all. Even when solutions are found, these generally lead to controversial discussions and disputes. In some cases, they can place the guiding principles of an organization in question and thereby provoke strong opposition with which the design thinking team must be prepared to cope. Therefore, in parallel to developing solutions, the team must foster and achieve respect, trust and thoughtfulness.

In practice, design thinking is increasingly establishing itself as a structured method for complex innovation and development processes. Well-known companies such as SAP, Deutsche Bank, UBS, Allianz, Volkswagen and Roche, among others, use these methods to develop new solutions (Zydra, 2014).

WHAT IS DESIGN THINKING?

What value can organizations expect to gain from using design thinking?

Innovative solutions: Organizations firstly expect innovative solutions from the method. Our experience shows that design thinking is suitable for both radical product, service and business model innovation, as well as the incremental adaptation of existing solutions. Through proven methods, teams are equipped to accomplish both.

Customer orientation and integration: The customer orientation and permanent integration of the client in the innovation process definitively demonstrate the high value of the design thinking method. On one hand, this helps ensure that results with which the client is in agreement are developed. On the other, this process also reduces project risks because testing with customers detects and eliminates incorrect assumptions and failures in design early.

The right questions: At first glance, this benefit seems almost banal. However, our experience shows that at the beginning of a project, the questions that should be answered through its course are often not clear. This is also completely normal when talking about innovation and future developments. Usually, the client has a rough idea regarding the area in which changes are needed. At this early stage in the process, however, the questions to be answered can be only vaguely expressed.

Design thinking helps us to express questions more precisely, or even completely revise them through its explorative, customer-oriented approach. In contrast to many other innovation and development methods, design thinking provides an explicit process step in which redefinition or more precise delineation of questions is conducted.

Divergent thinking: Management frequently orders employees to think outside the box. In opposition to this demand, organizations often have processes and structures in place to systematically reduce or even prevent exactly such risky activities. Design thinking is an appropriate instrument for promoting divergent thinking and ambiguity in projects in a controlled manner. The phase model sets a framework for explicit activities, such as the dark horse prototype (see page 44), that allow for engagement in new and sometimes eclectic perspectives on question statements and problems.

Tangible results: Design thinking fosters and requires communication through tangible prototypes. Building results and ideas into tangible forms is an integral part of the method, be it as product prototypes or service prototypes. This not only strengthens communication with the client, but also within the design team and the organisation. In many projects, instead of preparing time-consuming presentations, we have simply brought along prototypes to show to management. The reactions to these have been consistently positive. Instead of talking about possibilities and concepts, the discussion was immediately focused on concrete aspects of the proposed solution.

Agile approach: With its iterative approach, design thinking enables both quick delivery and the capability to modify outputs. Solutions and partial solutions are developed and tested in small steps. As a result, the process can be constantly managed and guided.

Risk reduction: Design thinking contributes to reduction of project and development risks. Through the agile process, failure can (and must) be encountered as part of the learning process—and early on. Faulty assumptions can be corrected early, which reduces risks in projects.

In this book we describe the design thinking method with defined processes and supporting techniques. Each design thinking project also initiates cognitive changes for the participants involved, so that in a broad sense, we can view design thinking from three perspectives (Schindlholzer, 2014).

1. Design thinking as culture, philosophy and mindset: A large number of those who practice design thinking consider it a culture and philosophy. Free of stiff models and processes, this group is primarily led by innovation principles in their search for products, services and business models. Often, this group cannot be identified with one clear understanding of the process.

2. Design thinking as a process: Design thinking as a process has the goal of making the method practical and manageable for a large number of people. The structuring of design thinking in frameworks, activities and techniques makes it possible for beginners to quickly and repeatedly put its principles into practice.

3. Design thinking as a toolkit: The third perspective sees design thinking as a toolkit. Techniques and learning can be drawn during the course of supplying design thinking projects to organizations.

THE MICRO-CYCLE—HOW DESIGN THINKERS WORK

OVERVIEW

The way design thinkers work is best described in comparison to traditional modes of working. The latter are defined by their rationalism, analytics and intellectualism. The empathetic, interpretive, and intuitive approach of design thinking stands in stark contrast. The following characteristics distinguish working in a design thinking team:

Practical

Design thinkers are generally practical and pragmatic. Instead of theoretically discussing concepts at length, design thinkers prefer to build prototypes and place them into reality. Through prototypes, abstract concepts and ideas are provided with a tangible, haptic expression. On the one hand, this supports communication with clients and stakeholders, and on the other, the literal tangibility of prototypes enable practical, pragmatic conversation and internal communication within the team. With the help of prototypes, uncertainties can be seen, discussed and, when necessary, changed. Prototypes make concepts palpable.

Reflection in action

Within organizations, many managers criticize the lack of agility in their teams. Design thinkers and the design thinking method embody action and agility. In the philosophy of a design thinker, exploration and reflection first happen through action. Action triggers trains of thought—practical action thus drives a diversification of thoughts. New insights are achieved, and creativity increased. Design thinking teams that incorporate this basic principle are agile and, in comparison to other teams, often extremely quick.

Tame and wicked problems

Both clearly structured and complex problems belong within the consideration set of design thinkers. However, the latter are often problems for which there appear to be neither a possible solution, nor a clear definition of the task or problem statement. To deal with complex, difficult situations, design thinkers have to learn to tackle ambiguity. Our experience has shown that this ambiguity poses serious challenges to project managers. Project managers typically learn to focus projects and reduce risks as early as possible. However, when the problem is recognized only in a vague sense, traditional project management tools fail.

Creativity

Within the context of design thinking, creativity can be understood in two ways. On the one side, design thinkers have the ability and aptitude to continually bring new ideas to the fore. Through this lens, creativity is related to a preference for the new over the old. The preference is for that which deviates and differentiates itself from the status quo. The goal is permanent renewal—on a continuous basis. On the other side, design thinkers experience creativity as a constructive and artistic work. Similar to artists, work goes beyond the purely technical development of the new. In addition, the experiential world is also pulled into the context of making (Reckwitz, 2012). Today, we often talk about precipitating a so-called "user experience"—that is, creating an experience for the user and customer in their use of products, services and business models. Design thinkers embody both aspects of creativity.

Visual and interactive

Design thinkers prefer to work visually. Instead of logging data and information into computer systems, they cover walls with Post-its. Instead of talking about abstract concepts, they quickly build prototypes.

Instead of writing only text on Post-its, they generate sketches that better imprint into memory. Visual communication not only makes conversations easier, but also helps design thinkers themselves put their ideas in order and think in new directions.

These characteristics are the basis of the central process of design thinking—the micro-cycle. The micro-cycle consists of five steps, which will be elaborated upon below.

1. Problem definition and redefinition
2. Need finding and synthesis
3. Ideation
4. Prototyping and storytelling
5. Testing

This cycle establishes the basis of the design thinking method. It is always repeatedly run through, in every single phase of a project. Repetition of the micro-cycle gets us closer to the goal step by step. It is an iterative approach.

The micro-cycle

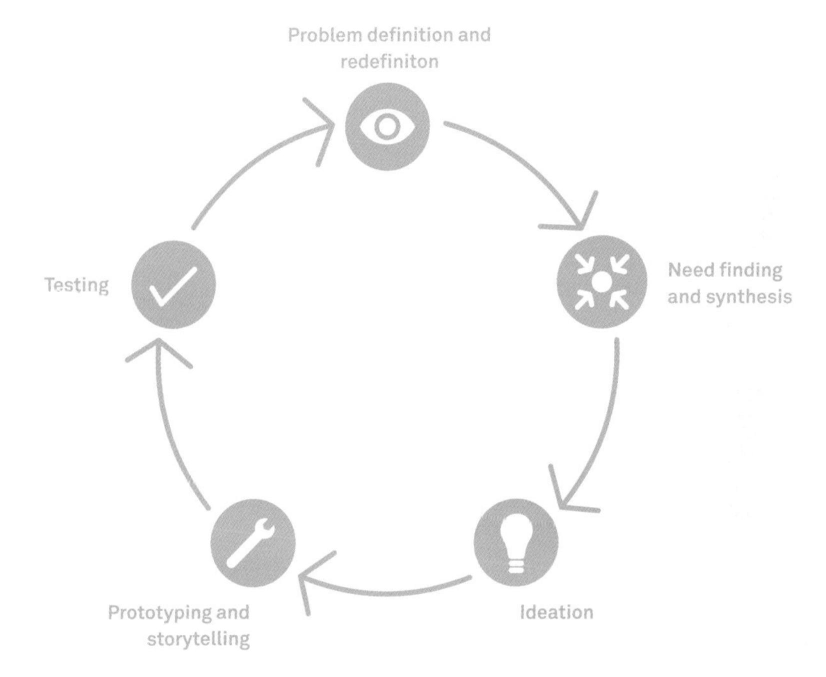

THE MICRO-CYCLE — HOW DESIGN THINKERS WORK

PROBLEM DEFINITION AND REDEFINITION

Every successful design thinking project begins with a good, goal-oriented question statement. Good ideas come from good questions!

But how can we formulate good questions? In design thinking, it is attempted to formulate questions as neutrally as possible. If a certain solution is focused upon too early, there is a risk of restricting the solution space in which we search for ideas. Instead, the question and problem definition should fuel the team's appetite for more information by evoking emotions, and curiosity at what might lie outside the box.

Most good questions contain four basic components:

Object: These are the products, services, strategies or business models that need to be newly taken into consideration or developed.

Audience: The question should speak to specific target audiences to whom the results of the project are expected to be relevant.

Framing conditions: The question should take framing conditions, which influence a given project at its outset and potentially in the future, into account. These framing conditions need to be considered in the solution.

Finished state: The final state of the object to be achieved should be defined.

An example of the formulation of a question statement within the context of design thinking could be:

How might this object be newly made so that, under the framing conditions A, B and C, the target audience can achieve end state X?

What makes design thinking unique is that newly defining the question is allowed, and even encouraged. Redefinition can take two primary courses:

Vertical division
Through project implementation, we quickly gain more knowledge than what we had at the onset of the project. This makes it possible to divide the initial question into subquestions to reduce complexity.

Horizontal adjustment
In some projects, conversations with clients make it apparent that the initial framing of the question does not address the client's needs. In this situation, design thinking enables us to reformulate the question upon the basis of data collected.

In the toolbox of this book, you can find numerous problem definition and redefinition tools.

THE MICRO-CYCLE — HOW DESIGN THINKERS WORK

NEED FINDING AND SYNTHESIS

Need finding is the design thinking phase in which the needs of clients, users, and other stakeholders, as well as insights (foundational realizations), are identified. The goal is to use the needs and insights to generate possibilities to newly develop or rework products, services and business models.

However, needs and insights cannot be equated with solutions. While the need of a client or stakeholder suggests a need for a change to an existing situation, a solution already presents a concrete implementation. Need finding concentrates on the identification of needs; solutions are built and developed only once we have entered the ideation and prototyping phases.

There are four techniques categorized under need finding.

Observation: Situations are observed in order to come to conclusions about customer behavior, product usage and context.

Interview: Conversations on concrete situations and future changes are had with customers, experts and stakeholders.

Engagement: The design thinking team places itself in the situation of a customer or stakeholder so that team members may experience the persona's position first-hand.

Benchmarking and business model exploration: In addition to investigating users, customers and additional stakeholders, design teams also focus on objective-oriented analyses of competitors and their business models. The objective is gathering knowledge to support the team's own solution space. In the benchmarking process, design thinkers search for situations comparable to the one on which they are working, which could help them to frame relevant questions.

These four techniques can again be found in the four steps of the need finding cycle (Becker & Patnaik, 1999):

1. *Framing and preparation:* Defining the research goals of the need finding process. Determining the customer and user groups to be investigated.
2. *Observation:* Carrying out and recording observations.
3. *Interviews and participation:* Carrying out conversations and taking part in concrete work, client and user processes.
4. *Synthesis:* Collecting, interpreting and analyzing information. Consolidated information helps us, in some cases, to adjust our initial questions (reframing).

For successful need finding, we must pay attention to the following points:

Planning: A good need finding process requires good planning before implementation. It is important not only to include clients and stakeholders, but also the techniques and questions to be used.

Environment and respect: Achieving an environment conducive to observation and carrying out interviews is crucial, especially in situations with confidentiality needs. Protecting the rights of individuals always takes precedence over all other goals.

Conversation partners: Conversation partners with a range of interests are ideal.

Synthesis concludes the need finding process. The data collected from and about clients must be intensified, built upon and expressed as learning. Techniques for synthesis, such as frameworks, support design thinkers in condensing new knowledge. A good synthesis, conducted over enough time, supports the ideation phase.

A large number of techniques are available for carrying out effective need finding and synthesis. Among these are camera studies and interviews. Many come from the field of ethnography and have been adapted for the purposes of design thinking over the past few decades.

Analysis of need finding interviews, UBS London (2015)

THE MICRO-CYCLE — HOW DESIGN THINKERS WORK

IDEATION

Ideation is the phase in which we build on need finding and synthesis by generating ideas for solutions. Brainstorming is definitely one of the best-known techniques. However, there are numerous additional techniques we can use to simultaneously draw on the data collected in the need finding phase and think in new directions.

Furthermore, there are basic principles that help to foster and implement the ideation phase.

Techniques: The techniques for finding ideas in design thinking are easy to use, even for beginners. Brilliant ideation results depend not only upon the use of techniques, but also the knowledge and experience of each individual participant with regard to the particular topic.

Customer integration: In many methods, not only design thinking, the customer or user is involved in the generation and evaluation of ideas. In design thinking, this is facilitated by the continuous production of prototypes representative of ideas and their direct testing with customers. This results in an evaluation process that in many cases awakens creativity in the testers. These often lead to new ideas that in turn lead to new prototypes. A virtuous circle results.

Curiosity: Curiosity is a characteristic that every team member should allow to thrive within him or herself. By being continually curious in trialling new situations, products or services—even when they may not appear promising at first glance—team members gain additional perspectives and further inspiration. When ideating, curiosity helps us engage with our teammates' perspectives, and to expand on them.

Reflection: Reflection is the ability of each team member to reconsider his or her own actions, or the actions of others, against the backdrop of the process as a whole. The examination of actions often leads to the recognition of fundamental principles, mechanisms or causes. The ability to reflect strengthens the ideation process through the recognition and advancement of thought and behavioral patterns.

It may seem astounding that design thinking does not immediately begin ideation at the outset of a project. The philosophy of design thinking is, however, that before we can ideate, we have to learn about the customer, user, and additional stakeholders. Ideation becomes useful only when we know what needs the customer may have.

THE MICRO-CYCLE — HOW DESIGN THINKERS WORK

PROTOTYPING

In this model of design thinking, prototyping is neither a plaything nor a craft session for employees. Prototyping is a technique in which real products, services and business models are simulated in the form of prototypes. The goal is that a development department directly affiliated with the design thinking project can realize these prototypes.

As a phase in design thinking, prototyping describes the concretization of ideas. As an activity, it is the transformation of pure ideas into tangible objects. The prototype is the object that emerges — the concrete instance of an idea.

As a phase and an activity, prototyping connects many users (Ulrich & Eppinger, 1995).

In-team communication medium: Prototypes allow discussions to move from the conceptual and abstract to a concrete, tangible, and thus verifiable level. While conversations over concepts often lead to misunderstandings due to a lack of tangibility of the object under discussion, prototypes already offer a bridge from concept to product early in the project process. Using this tool, the design thinking team can articulate themselves and achieve mutual understanding.

Uncovering previously unforeseen problems and challenges: Both the prototyping activity and prototypes help us to quickly identify problems and challenges. Through the tangibility of prototypes and the work required to generate them, new challenges come to light. Some of these would be otherwise discovered only through lengthy discussion and deliberation; others would likely never be revealed until too late.

Raising new questions: The involvement and debate that accompanies the process of building prototypes forces new questions to arise. The phrase "learning by doing" is completely applicable here.

Reducing project risks: The early building and testing of functions leads to the early recognition of new needs, challenges, and problems. Especially during the early phases of a project, this knowledge helps to reduce project risks — or even prevent them from ever arising.

Testing assumptions: Particularly in early phases, the project team frequently works upon the basis of assumptions rooted in their own experiences. With the help of prototypes, assumptions can be tested on customers and revised early, if needed.

Refinement: Prototypes help us to more rapidly expand and refine our ideas and conceptions.

Customer feedback-based decisions: Prototypes tested with customers help us to make decisions about next steps on the basis of concrete data.

Trust and gut feeling: Finally, prototypes develop trust in the team's own work and a feeling for the direction of further development.

"Prototyping is problem solving. It's a culture and a language. You can prototype just about anything — a new product or service, or a special promotion. What counts is moving the ball forward, achieving some part of your goal. Not wasting time."
(Kelley & Littman, 2001)

The six "golden rules" of prototyping have emerged from numerous completed projects. Alexander Grots summarizes them below:

One question—one prototype: The team should try to develop the prototype in response to one single question. The fundamental principle of "stay focused" is at the center here.

Don't perfect: Prototypes are prototypes, not products or services (yet). The team should focus only on the characteristics and level of detail that are relevant at that moment—and not more (Brown, 2009)! In early phases of the project, this means focusing on predetermined functionalities; later in the project prototypes should be riper and thus able to combine multiple functionalities.

Don't fall in love: Team members or the team as a whole should not "fall in love" with their prototype (Uebernickel & Brenner, 2015). Results of testing with customers can show that ideas do not work. In such cases, the team also has to be able to let go of failed ideas and prototypes. It can be difficult to let go, especially when lots of time and energy has gone into a prototype. In these situations, one is often inclined only to partially redevelop the existing solution, instead of starting again from the beginning.

Push boundaries: Prototypes should be used to try to transcend existing boundaries (technology, usability etc.). The team should not take statements like "that won't work, we already tried that" as demotivating, but rather as provocation to overcome hurdles.

Provoke to convince: Prototypes should not be familiar, but provocative to the tester. New developments tend not to inspire enthusiasm at first. Therefore, the team should be brave! For our project with a soccer organization, we focused on aspects of soccer player registration. The existing system stipulated that soccer players always be registered by their soccer club. Nevertheless, our team intentionally provoked by building a prototype system in which players could register other players without the involvement of the soccer club. As expected, this prototype provoked some stakeholders—and we thus gained valuable insights, especially regarding the behavior of young soccer players. For this, the provoking approach was perhaps the most effective.

Break the rules: Teams should knowingly break rules in order to try out new, alternative scenarios. These rules can be externally imposed or may have developed internally over time.

Numerous techniques are available for prototyping: Their variety reaches from rudimentary prototypes using paper, cardboard and Post-its to complex prototypes made of electronic components, mechanical parts or software.

Fundamentals: At the beginning of the design thinking process, the development of low-resolution prototypes is useful. They are cheap, easy to build, and quick to complete. Moreover, the design thinker does not run the risk of falling too deeply in love with his or her own work (Uebernickel & Brenner, 2015). In the later phases of a design thinking project, more detailed, high-resolution prototypes are built. These are generally demanding in terms of content and technology, therefore their materialization takes longer. The goal of the high-resolution prototype is to offer a template to guide the development of a real product or service, and offer the customer a real working simulation of the solution generated.

1

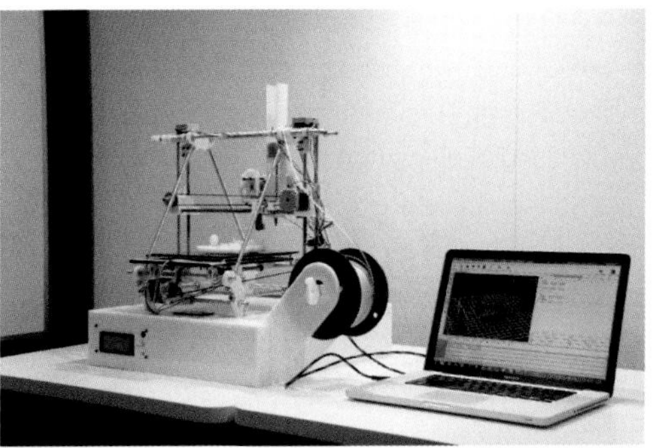

2

3

1 Putting paper bikes together at Radicand Labs, Palo Alto (2014)
2 Prototype of a 3D printer for electronic components (2014)
3 Testing a software prototype for a financial service provider in Cali, Colombia (2013)

THE MICRO-CYCLE — HOW DESIGN THINKERS WORK

TESTING

In the testing phase, ideas and prototypes are tried out and evaluated by customers and other stakeholders. This means that the previously made assumptions and initial approaches to solutions are tested for validity. Feedback from tests is optimal when it inspires team members and pushes them to develop new ideas—without them being strongly attached to the prototype—whilst simultaneously helping them to understand the problem statement and thus learn how a solution might be implemented. Testing assumptions and filtering out misassumptions are important in the start phase of design thinking projects. Through these tests, the team learns about important aspects of the problem and solution spaces. Reflecting on this as a team can lead to a better solution.

Testing serves three purposes:

Inspiration: Prototypes should inspire not only the team, but also customers and testers to think about further improvements, further themes or new question statements. An example was a reconception of the assumption of how cars might be received for repairs in large cities such as Shanghai and Beijing. The team drafted a new concept in which cars were received in shops with a design similar to Apple Stores. During testing, customers were presented with first architectural prototypes and their inspiration was awakened. The feedback sessions helped them to improve results in further iterations, and find their own interpretation for the automobile manufacturer.

Evaluation: Prototypes are assessed with reference to key criteria stipulated in advance, such as application speed, user friendliness, shelf life and others. The evaluation typically takes place within the team and not (yet) with users and customers. For example, in 2005 we had the opportunity to work with a product similar to an iPad. Early in the project, we tested the maximum acceptable weight of the product with users. Thereafter we could always refer back to this tested weight limit when evaluating newly constructed technical prototypes for weight.

Validation: After assessing prototypes on use and function, we test whether or not the characteristics evaluated actually add value for the party the prototype addresses. In numerous situations, validation calls for us to directly interact with customers on location. In the aforementioned project for a soccer organization, we traveled around several countries in order to test new approaches to managing soccer players. The experiences in countries such as Ethiopia and Uruguay were absolutely necessary for the team to develop a feeling for the views and ideas of resident soccer associations and players.

As with need finding, certain framing conditions apply to testing:

Planning: Testing prioritizes the tester. Appointments for conversations and focus groups need to be planned and organized early, and kept. Teams can seldom expect testers to be available at their convenience. Although business-to-consumer questions and testing can often take place on the sidewalk, experience has shown that planned testing often yields more productive results, and leaves less to chance.

Politeness and respect: Test participants should be treated with respect. In addition to professional responses to disagreeable results, this involves protecting individuals' personal rights.

Well-known testing techniques include "consumer clinics" and "usability testing". However, need finding techniques, such as interviews, can also be employed.

1

2

3

1 Paper Bike Challenge at Stanford University, Palo Alto California — paper bike testing (2014)
2 Prototype testing at Tongji University, Shanghai (2015)
3 Testing a software prototype for a financial services provider in Cali, Colombia

THE MACRO-PROCESS: THE 7+1 PHASES OF DESIGN THINKING

OVERVIEW

The macro-process structures the design thinking approach and is worked through once over the course of the project. The macro-process is spread over seven steps, each of which contains one or more iterations of the microcycle (see p. 24). The macro-process is separated into two stages: the diverging stage in the first half of the project, and the converging stage during the second.

The diverging stage has the goal of systematically expanding the conceivable space in which a question statement can be formulated. The purpose is encouraging the team to widen the solution space. During the divergent stage, as many ideas as possible are generated, constructed as prototypes, and tested with users (Uebernickel & Brenner, 2015). Even when an idea or solution is very positively received by users, the design thinking team nevertheless tries to continually expand the solution space.

The focus of the converging stage is evaluation of all the results produced during the diverging stage with the intent of building a (single) final prototype. Both positive and negative results from the prototype testing of the preceding stage are evaluated. The aspects considered relevant are combined in a comprehensive final prototype.

The entire macro-process of divergent and convergent stages is divided into seven steps:

Design space exploration
Design space exploration is the starting point for a design thinking project. In this step, the project plan is not only laid out, but early explorative need finding is conducted with customers and other stakeholders.

Critical function prototype / critical experience prototype
During the critical function and critical experience prototype phases, potential solutions for the functions considered critical are built into prototypes. Critical function prototypes are neither final nor comprehensive, but instead propose individual pieces of a wider potential solution. The goal is to make single elements of a potential solution testable as quickly as possible. Critical experience prototypes concentrate on the perceptions of customers and users, rather than a function. As with all prototypes, the team should use both critical function and critical experience prototypes as an opportunity to experiment, gain new experiences, and learn.

Dark horse prototype
The goal of the dark horse prototype is to encourage the team to become even more open to new approaches, ideas and potential solutions. While the critical function prototype sees the team taking the project's framing conditions into account, the dark horse prototype calls for the team to ignore or perhaps even knowingly pursue the opposite of what the context elicits. Specialized reframing techniques help the team to widen the boundaries within which they search for questions and answers. Through this process, teams are able to consider solutions unable to receive attention in the context typical of the client organization. These tensions result in dark horse prototypes typically generating the most controversial discussions during testing—dark horse prototypes propose solutions perceived as having low probability or feasibility. However, it is exactly this tension within both the team and testers that creates the potential for new discoveries. As the project is still in the diverging stage at this point, numerous prototypes continue to be built.

Funky prototype (integrated prototype)
Funky prototypes form the bridge between the diverging and converging stages of a project. The solution space for ideas and solutions continues to be widened, however the primary objective of

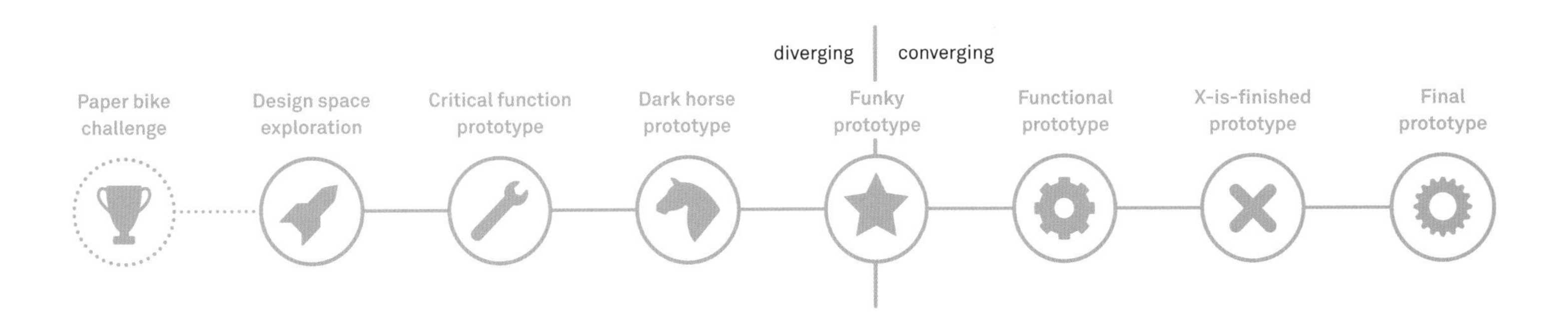

the funky phase is the combination of the divergent phase's most successful solutions and interesting learning. The focus of the funky phase is the functionality rather than the aesthetics of a solution. Funky prototypes are higher resolution than the prototypes generated in preceding phases—and thus more demanding in terms of their construction. In practice, it has proved useful to call this step the "integrated prototype" as employees find it easier to connect with this term.

Functional prototype

Functional prototypes should already lay the basis for future extrapolation into complete final prototypes. The functional prototype forces the team to examine the technical and managerial details of a solution and address questions that may have remained unexplored up until this point. In many cases, this is the point at which all the functions of the eventual final prototype come together for the first time.

X-is-finished prototype

The x-is-finished prototype is the last step before the final prototype. The goal is to completely build one single function or component of the finished prototype. The component selected should not be trivial, and must contain complexity. The main purpose of this phase is to assess the time, resources, and effort required to generate a prototype of the function or component so this concrete data can be used in further project planning.

Final prototype

The final prototype presents the result of the design thinking project. It contains the critical functions of the new solution. The final prototype should be as realistic as possible and illustrate all functions to customers and users.

"Your brain is a synthesis machine—embrace new combinations (and divergent thinking)"

(Georg Kress, 2014)

The final prototype serves as the basis for developing the set of specifications needed to put the solution into production.

Note: the prototyping phases are all described above with reference to a single prototype—however multiple prototypes of the same type (e.g. funky/integrated prototype) are typically built during each phase!

Depending on the experience level of a design thinking team, the length or intensity of a given phase can always be adapted according to the project and its requirements. Additionally, the macro-process (see p. 36) determines where work in the micro-cycle (see p. 24) should be focused. During the first steps (design space exploration, critical function and dark horse prototypes), the focus is squarely on identifying customer needs and quick prototyping. It is essential to review and test as many assumptions as possible. In later steps, however, more time is taken to produce and test higher-resolution prototypes. At the same time, the time for analyzing customer needs is reduced.

At every step of the macro-process, multiple steps of the micro-cycle are typically under way. Depending on the design thinking team's experience, members are able to assess which steps are really necessary in a given situation, and which ones can be skipped. However, we recommend that beginners view the macro-process and micro-cycle as obligatory sets of rules.

This said—one step is still missing!

The paper bike...

In academia, every design thinking project starts with a paper bike challenge at Stanford University in Palo Alto, California. The teams must build a working vehicle, capable of carrying at least one person, using only paper, cardboard, and a maximum 500 grams of miscellaneous material such as glue or metal. This paper bike must survive various games and competitions.

The paper bike challenge is a big warm-up for the students with the primary goal of: Have fun! The focus is on the experience of accomplishing something with one's own two hands. At a business school such as the University of St. Gallen, it can be bizarre to see students roaming the halls with hammer drills and saws. However, over time even the university directorship has grown used to this scene.

Paper bike challenge, Stanford University, Palo Alto, California (2013, 2014)

THE MACRO-PROCESS: THE 7+1 PHASES OF DESIGN THINKING

DESIGN SPACE EXPLORATION

The time has now come for the team to hit the streets.

Design thinking projects always start with exploration of the design space. The design space describes the direct environment of the question statement provided. To become familiar with this, the design team must carry out desk research on the one hand, while carrying out exploratory conversations with customers and other relevant stakeholders on the other. This serves as the first information from the field, which must not only be collected, but also structured (need finding and synthesis).

The following, among others, belong to a good design space exploration:

Customers: The design thinking team identifies the probable or known customers and customer groups. In doing so, the team collects the first leads regarding potential classifications and groupings. These help them to identify and recruit the right people for interviews and testing in the subsequent need finding and synthesis phases. The same process applies to users. A common method for gaining an overview and transparency is called the stakeholder map.

Stakeholders: In addition to the user there are often additional parties that are relevant to a project—these are called stakeholders. Stakeholders are identified and grouped just as customers and users are. Information should be collected in a way that allows for later changes and elaborations. Stakeholders can be individuals, or groups and organizations in which the individual people are not known.

What does the team know, and what doesn't it know yet?

In a design thinking project, there must be transparency from the very beginning on what the team does and does not know for answering the question statement. This clarity provides direction on which knowledge must be acquired as soon as possible. The team must maintain an inventory list of knowledge throughout the course of the project.

Benchmarks

Benchmarks refer to comparable situations in other industries or sectors. For example, a team aiming to optimize an operating room might use Formula 1 as a benchmark. Both fields share a focus on the highest level of precision and speed, among other things. Mistakes would almost certainly have catastrophic results for a human being. Through benchmarks, the team can study processes, draw parallels and transfer them into their own question statement.

Patents and further information sources

Patent databanks and further sources of information, such as technology journals, are also important sources of knowledge for the design space exploration phase. The team should continuously monitor these as sources of new trends, knowledge or other interesting findings to benefit their own project.

Business models and start-up screening
Keeping a vigilant eye on new business models and young enterprises (start-ups) is part of design space exploration. For this reason, portals such as kickstarter.com or indiegogo.com should be on every design thinking team's surveillance list. Other useful websites include angel.co and crunchbase.com.

Trend scouting
Trends are a useful source for enriching the design space. Trendwatching.com and psfk.com are well-known sources. A number of providers offer studies on trends related to all manner of circumstances and technologies. Design thinking teams should regularly search for trends related to their specific design space.

Design space exploration,
Johannesburg, South Africa (2014)

THE MACRO-PROCESS: THE 7+1 PHASES OF DESIGN THINKING

CRITICAL FUNCTION PROTOTYPE AND CRITICAL EXPERIENCE PROTOTYPE

It is always difficult to start! This is also often true for design thinking teams. The question statements developed and their surrounding environments are often so complex that it can be difficult to figure out where to begin.

In order to make this complexity manageable, the critical function and critical experience phases have the goal of developing prototypes for isolated singular functions or experiences. Ideally, the selected functions should be chosen to feature novelty and surprise value from the perspective of customers and testers. The goal is not to merely get positive feedback from the user or customer. In some cases, it is worth being provocative in order to concretely define the boundaries of what is possible and acceptable.

Low-resolution critical function and critical experience prototypes are useful for demonstrating ideas. Using this approach, the team can quickly produce numerous prototypes and take them to customers and users for testing.

But what is the difference between critical function prototypes and critical experience prototypes?

Critical function prototype (CFP)

The CFP is a physical product or service provision system that, in the form of a prototype, allows customers to test a given function or component. It is definitely not always easy to specify and build just one function. If this is not possible, teams should try to at least reduce the number of functions present in their prototype to a minimum.

Critical experience prototype (CEP)

The CEP should demonstrate "just enough" of an experience to the tester to enable the design thinking team to extract learning. An example of a CEP is the simulation of a self-driving vehicle. For this, the team could build a cabin in a lab. The cabin might contain real car seats and multiple screens displaying video footage of a real driving scenario. This type of prototype would be useful for very simply testing the experience of a self-driving vehicle. Of course, in this day and age there are more sophisticated possibilities, for example the insertion of simulated turns and curves. However, the resolution described above is sufficient when the objective is to acquire initial learning for the project.

In a typical design thinking project, numerous prototypes are developed during this phase. It is important that they all lead the team to build and craft.

"Design a component or function that makes a difference."

1

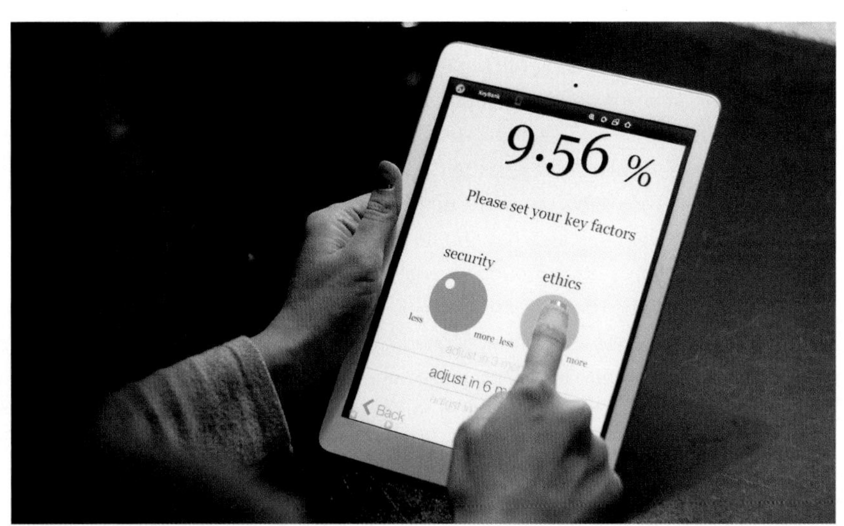

2

1 Prototype of a financial product called "in the box" (2013)

2 Prototype of an interest calculator based on personal customer preferences (2014)

THE MACRO-PROCESS: THE 7+1 PHASES OF DESIGN THINKING

DARK HORSE PROTOTYPE

The name “dark horse” comes from a horse racing anecdote. The dark horse is the horse on which no one bets—yet which defies expectations and wins. This is apt for dark horse prototypes. Dark horse prototypes are abnormal, even to the point of being comical. They depart from the limiting aspects of their given realities. They should intentionally topple consensus around emergent solutions and introduce uncertainty. Why? Practical experience has shown that consensus-oriented solutions are often only incremental modifications of existing solutions. In contrast, solutions that polarize customers or markets—or even appear absolutely unfeasible—have the potential to become radical innovations. Even if these prototypes cannot (yet) be realized, single elements can perhaps come to fruition. It is critical that the design thinking team does not shy away from ideas due to their initial apparent impossibility.

In many cases the dark horse prototype phase is marked by numerous reframing activities (see toolbox). In reframing, numerous question statements are considered from multiple perspectives and rearticulated in order to open expansive new perspectives on an issue.

Similar to the CFP and CEP phase, the prototypes built are almost exclusively low-resolution. They are simple, fast and cheap so that testing can be quickly initiated. Using this approach, failures are easy to cope with and benefit the team's learning process.

According to our experience, it can be difficult for dark horse prototypes to gain traction in traditional, hierarchical companies. The conscious departure from reality often leads management to merely laugh at projects. On a few occasions, they have even led management to consider abandoning a project. This phase is often connected to a perception that the project has too much scope for messing around and a lack of focused work. Such an opinion should be reassessed because it misses the point of the dark horse phase. Exactly this playfulness and freedom is what the organization needs to overcome performance barriers and develop completely new solutions. In numerous projects, particularly during the critical function and critical experience phase, team members often focus primarily on their client organization's existing business model. Only in the dark horse prototype phase are they able to overcome their own minds' limits, and consider new business models—if need be contrary to those of the existing business.

“Sometimes, when you have a crazy idea, it's worth making it happen.”
(Hilary Hahn, 2015)

1

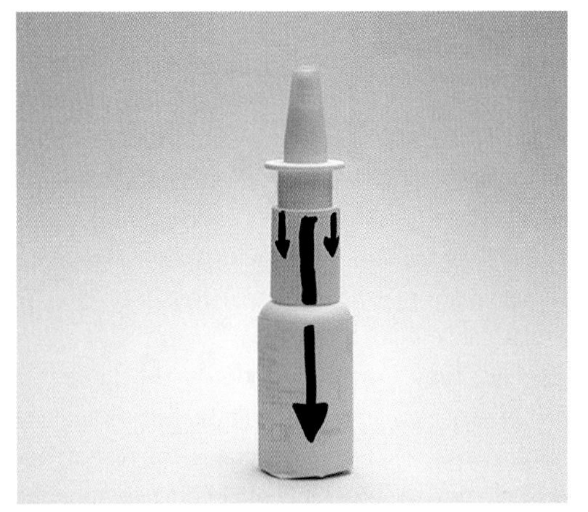

2

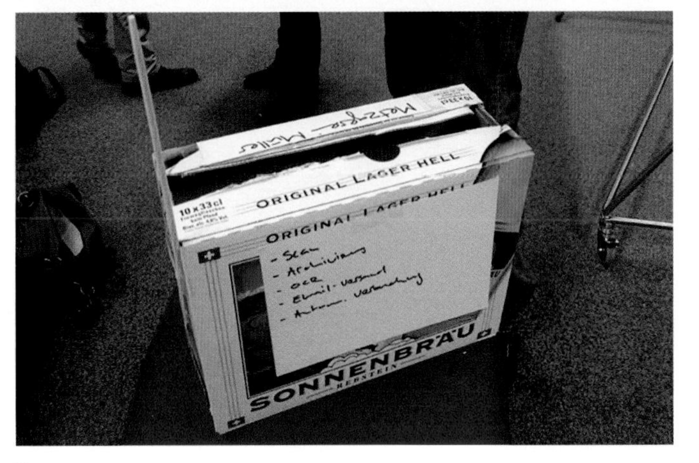

3

1 Dark horse prototype for lane-dependent road user fees (2011)

2 Dark horse prototype of a nasal spray bottle that vacuums mucus (2014)

3 Dark horse prototype for a device that fully automatically scans, interprets, and files records and documents (2012)

THE MACRO-PROCESS: THE 7+1 PHASES OF DESIGN THINKING

FUNKY PROTOTYPE (INTEGRATED PROTOTYPE)

Funky prototypes (also called integrated prototypes) help us pose critical questions related to production and delivery. They combine the most successful and interesting solutions from the critical function, critical experience and dark horse prototype phases into new and integrated solutions. Through the combination of different solutions, this phase also becomes a high point in terms of the number and diversity of ideas and prototypes. Funky prototypes thereby constitute the bridge between the diverging and converging stages of the micro-cycle. An important activity in this phase is the determination of which previously failed prototypes should be excluded from further consideration.

Funky prototypes are still far from being haptically and aesthetically finished. However, this category of prototype should provide a sketch of the essential functions and elements of a potential solution

to the problem originally defined. In this phase, the focus of the micro-cycle is above all on building and testing instead of additional need finding.

From the perspective of the design thinking team, this phase is demanding because decisions need to be made on which prototypes to further develop, and which to abandon. At this stage, discussions both in the team and with clients are typically energetic, as team members and other stakeholders grapple with the best solutions.

In our experience, the number of prototypes reduces slightly in this phase compared with previous phases, the reason being the increasing intensity of building complex prototypes, as well as the continuous integration of different individual prototypes into a single solution.

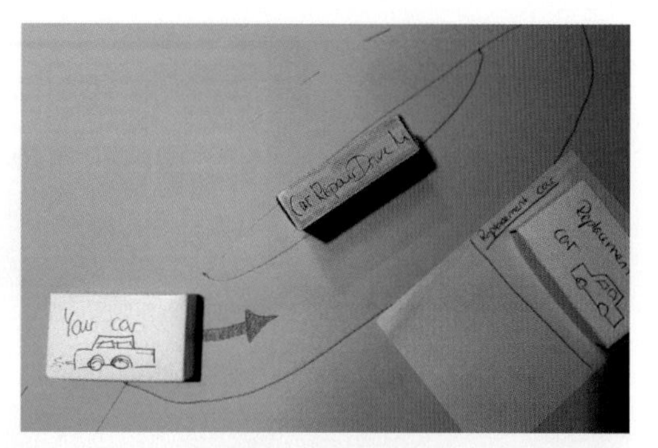

1

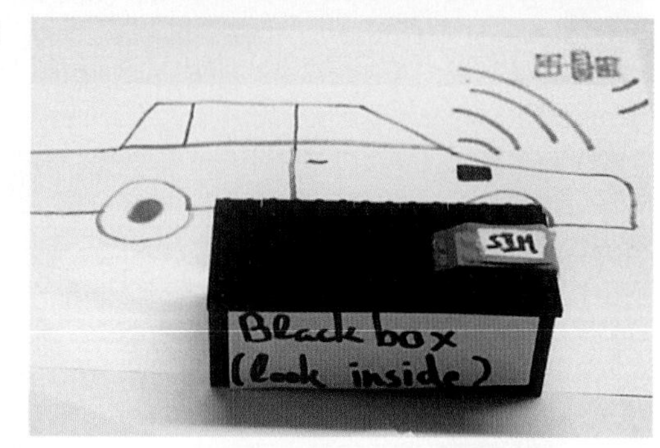

2

1 Integrated prototype demonstrating a new way for cars to enter repair shops (2013)
2 Integrated prototype of a black box for cars that reads parameters for opening/closing as well as starting/stopping (2012)

FUNCTIONAL PROTOTYPE

With the functional prototype, the requirements for ultimate solutions and their scope are set out. The functional prototype makes an extrapolation of the final prototype possible. The comparatively high degree of elaboration allows the team to estimate the effort required for implementation, and discuss it with the client.

Furthermore, the high degree of detail makes new testing methods possible. Individual functions can be evaluated and validated in a comprehensive test. Therefore, this phase is often also the starting gun for integrating additional individuals and companies into the preparation of an ultimate solution. For example, if the team needs to produce software or complex electronic modules, software developers may be brought into the project at this point.

Functional prototype of a steam-spray nozzle (2013)

THE MACRO-PROCESS: THE 7+1 PHASES OF DESIGN THINKING

X-IS-FINISHED PROTOTYPE

With the x-is-finished prototype, the team can estimate the level of effort required for the implementation of the final prototype. The goal is to develop and complete a selected function or component of the final prototype. The team should ensure that the component selected is sufficiently complex and challenging. After completing the x-is-finished prototype, the team can again assess which functions and features should be transferred into the final prototype.

Clients of design thinking projects tend to overload the list of components and functions to be implemented. At the same time, they see interruptions to scheduled everyday appointments within their organizations. The x-is-finished prototype therefore offers a good opportunity to review whether or not all the functions and components proposed can be implemented in the remaining time window. If not, the list of functions and components should be shortened.

NOTES / SKETCHES

THE MACRO-PROCESS: THE 7+1 PHASES OF DESIGN THINKING

FINAL PROTOTYPE

Finished! The final prototype presents the design thinking team's solution to the question statement they developed. The quality of the result should be so high that customers, users and other stakeholders can give sound, comprehensive feedback. The final prototype can be a product, a service, a strategy or a business model.

After a successful final test, the final prototype also serves as the basis for setting specifications for further development and implementation in the organization.

1

2

1 3D printer for electronic components (2014)

2 Final prototype of a family welcome box, to be given to the family after the birth of a child, as is customary in Finland (2013)

THE DESIGN THINKING TOOLBOX

The design thinking toolbox is a collection of methods that provides a systematic and structured approach to solving problems.

Beginners can use the methods described in this chapter as a guide for successfully working through a design thinking project. The methods help us to grasp the complexity of a project and maintain a focus on the essential factors.

In contrast, those who already have experience using design thinking methods can use the collection of tools here as a source of inspiration in their work. Elements of these methods can be modified, and some predefined steps can be skipped, carried out multiple times, or adapted. The toolbox is not a static ordinance.

In order to make the toolbox more manageable, it is organized as if it would follow the micro-cycle through the course of an idealized design thinking project:

Problem definition: A successful design thinking project always begins with a good question. The methods in the problem definition phase show how to develop a good question, and what factors are important to pay attention to during its formulation.

Need finding and synthesis: Methods of need finding and synthesis help the design thinker to understand the problem and extract learning or direction for the next steps in the project. During observation, the design thinker deepens his or her understanding of all the customers, users and stakeholders within the scope of the question at hand—regardless of whether it is a business-to-customer (B2C) question or a business-to-business (B2B) question.

Ideation: Ideation methods help the design thinker to optimally activate his or her creative potential.

Prototyping: Prototyping methods should make ideas visible and tangible. Translating product ideas into prototypes is intuitive. In this chapter, however, we also share approaches to building prototypes for services and business models.

Testing: Testing with customers, users and stakeholders is needed to evaluate the ideas the design thinking team proposes. In this section, we present selected methods that can be used in B2C and B2B contexts.

To supplement the methods in the micro-cycle, three additional chapters that cover techniques have been integrated into the design thinking toolbox:

Project setup: Every project requires structure and planning. Design thinking projects are no exception. In this phase, methods help us to set up a project, but also to monitor its progress.

Warm-ups: Warm-ups are little exercises and games, that help us to wake our bodies and minds. Typically, warm-ups are intensively used in advance of an ideation session.

Feedback: Design thinking teams live on feedback. This chapter demonstrates a few techniques for structured feedback.

The design thinking toolbox is comparable to existing industry norms, for example VDI 2223 (methodologically-devised technical products). In addition to a process model showing concrete steps, it offers techniques that enable a given step to be completed as effectively and efficiently as possible (Association of German Engineers, 2004).

Some of the methods described herein can be placed in more than one category. In such cases, the method is placed in the category used most frequently.

balls
ALUFOLIE
surprise

SPIELZEUG
colored paper

ROLLEN&GROBES

duct tape
creped paper

PFEIFFENPUTZER & KABELBINDER
BÄNDER
EINWEG-BESTECK

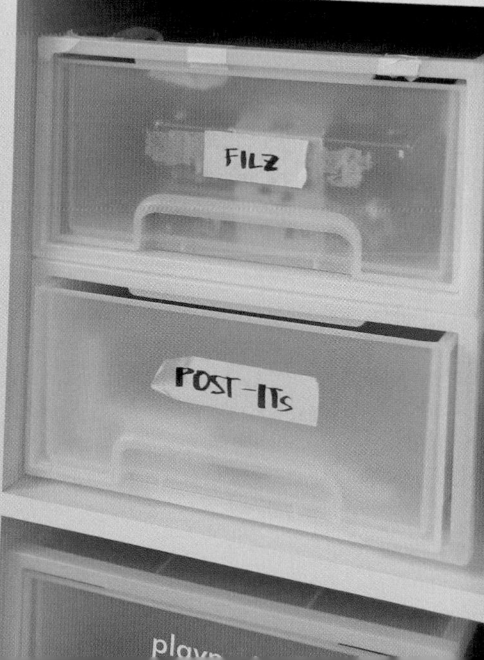
FILZ
POST-ITs

THE DESIGN THINKING CODE

The success of design thinking projects is heavily dependent upon the trust of customers and other conversation partners. Without engagement and open collaboration with people outside the team, design thinking would fail as a customer-oriented method of development and innovation.

Design thinking teams find themselves in contact with potential users and customers as often as possible. The goal is to analyze and understand people and their behavior, and to integrate this into the client organization's thinking. Often, this contact not only takes place in the design thinking team's home country, but requires on-site investigations of other cultures. For example, in a project for a pharmaceutical company, we carried out research in the target market of South Africa as well as countries in Europe, the Americas, and Asia. The design thinking team interviewed people living in affluence, as well as people in Soweto, Johannesburg, South Africa living in very poor conditions. Without the help of these people, the project would not have been possible.

This openness and willingness to collaborate extensively depends upon trust in our work. Without trust, these conversations would simply not happen. We would not have the opportunity, for example, to take the customer and user photographs and videos so central to the design thinking process. Trust does not just happen—individuals and enterprises must respectfully develop and nurture it. One form with which to nurture trust is the strict and consistent adherence to rules that apply regardless of cultural circle, country, gender, age, educational level, philosophy and so forth. We call these rules the design thinking code.

The design thinking code is based on the ethical rules used for international market research established by the European Society for Opinion and Market Research (ESOMAR, 2007). The rules in the design thinking code are normative and provide the basis for all exchanges. Like those of the ESOMAR standard, the design thinking code is a minimum standard for the behavior and actions of design thinkers. When national and international laws are taken into consideration, there are stricter rules and regulations under a number of circumstances.

The purpose of the design thinking code is (in alignment with ESOMAR 2007):

- Providing uniform norms and established rules for design thinkers that serve to regulate ethical interactions with people and collected data;
- Enabling the responsible treatment of people and society by transparently communicating norms and rules;
- Illustrating particular responsibilities with regard to children and young people when conducting research and testing, in addition to delineating responsible conduct within the team itself.

In alignment with the international standards for market researchers (ESOMAR, 2007), design thinkers also adhere to the following basic principles:

Jurisprudence

Design thinkers must know and adhere to national and international laws. The given laws can, depending on the circumstances, lead to a tightening of the design thinking code.

Behavior
Every design thinker must fundamentally adhere to standards of ethical behavior. The behavior of one individual cannot be allowed to disadvantage the whole design thinking community.

Children and young people
Great care must be taken during research (need finding) and solution testing with children and young people. This applies in terms of how we behave with children and youth, as well as data collected.

Voluntariness
Participation in interviews and observation is voluntary! The information presented to participants must be appropriately prepared in understandable language.

Personal rights
The personal rights of private individuals should be respected and adhered to. Infringing upon or compromising the rights of private individuals should be avoided, as should exposing the identity of participants.

Breach of purpose
Design thinkers may never use the data they collect for any purpose other than those communicated to the participant, and those for which the participant has explicitly given his or her permission.

Transparency and objectivity
Design thinkers must ensure that all activities and projects are transparently and objectively conceptualized, implemented and documented from beginning to end.

The code is also associated with interaction with data in computer and information systems. The undisclosed saving of data itself can already lead to a breach of data privacy rights. Therefore, at a minimum, the following aspects of law and ethics must be considered. Some individual cases will certainly raise additional issues.

Anonymity
Data from interviews, observations or participation in field visits must remain anonymous unless other arrangements have been directly made with participants. This means that information that could lead to a participant being identified must be removed from the dataset. This includes names, addresses, or identification numbers from insurance agencies or personal identification cards. Only anonymous data can serve as the basis for further work. For large datasets, removing identifying characteristics takes a substantial amount of time; nevertheless, this work is crucial.

Safeguarding data (access protection)
The protection of data from third-party access, even if anonymized, is a complex issue. In our digitized world, there are so many points of entry into computer systems that an unskilled user is quickly overwhelmed by the steps one must take to protect data. Nevertheless, some of the issues to be taken into consideration are outlined below:

Hard drive encryption: For design thinkers who travel frequently, computers can be easily left on trains, or even stolen. Encryption of data on hard drives offers at least a first hurdle to intruders if/when they attempt to access or otherwise take advantage of stored data.

E-Mails: Unfortunately, e-mails are not encrypted or secured frequently nearly enough in today's world. However, when sending personal and confidential data, it is necessary to draw upon encryption standards. We know that this does not always happen for purely practical and pragmatic reasons. However, when the possibility exists, one should always try to use encryption.

USB sticks: Of course it is practical to transport data from point A to point B with USB sticks. However, we must be aware of the fact that USB sticks are easily lost. The finder of a lost USB stick then has unrestricted access to the data it holds. If you choose to use USB sticks, encrypt them! Then at least a basic level of protection is achieved.

Data destruction

Unfortunately most design thinkers do not think about the destruction of data. However, the question quickly arises after the successful completion of a project: "Now what do we do with the data?" Easy—as long as further use has not been arranged with participants, the data should be irreversibly destroyed!

"Ethics is a process of learning—not a process of obedience."

(M Lab)

Street interview, Shanghai, China (2015)

THE TEAM

Good teams are key to the success of design thinking. Only when each team member is ready to give their best, and the interplay between teammates is in sync, can teams gain the motivation, momentum, and drive necessary for developing creative and innovative solutions. So what characterizes a good team? What kinds of people do they need, and what should one pay attention to?

Mark Schar, a colleague at Stanford University, once succinctly stated that everything one needs to learn about good teams can be seen on television. He delivered examples from the film series *Power Rangers* and *Star Trek*; Captain Kirk, who pays constant attention to the big picture and makes hard choices in tough situations, Commander Spock, who constantly questions (primarily human) behavior, and "Number One", who is not only first officer—but also valued for her logical thinking and analytical capabilities. These are just a few of the characters from which one can learn what makes exceptional teams.

Design thinking works in exactly the same way. It is not about looking for a certain kind of person, such as in the motto: "We need creatives." No—outstanding teams are comprised of many different types of people. The interplay and mix of competencies is critical. In design thinking, successful team composition depends on three factors: personality type, multidisciplinarity and T-profile (see next page). Moreover, it is the way in which these individuals uniquely complement one another and build synergies that makes them part of a great team.

Personality type

A person's personality type is not an indicator of "good" or "bad", but rather a support in building an optimal team with diverging characteristics. According to McGregor (2006), "the right mix at the right time" is decisive. For a good team, it is important not only for members to have different educational backgrounds (multidisciplinarity), but also different personality types. The interplay between introverts and extroverts, or thinkers and communicators, makes for a good team.

In student teams, we use the Teamology test to segment potential team members into 16 groups. The primary goal is to put together groups of people with as many different characteristics as possible – analogous to the *Star Trek* example. In organizations, this method often quickly reaches its limits, as the rights of the individual are naturally of utmost concern. However, open discussion with human resources departments and employee associations almost always leads to constructive results.

Multidisciplinarity

Multidisciplinarity means that in design thinking teams, we try to integrate as many educational backgrounds as possible—for example, engineering, computer science, business administration, psychology, design, architecture, and so forth. Team members' expert knowledge and also their different problem-solving approaches, offers teams the opportunity to establish themselves. Together such teams are stronger than the sum of their parts.

Companies often try to place employees from different departments into teams. In some projects, financial controllers have worked together with their colleagues from R&D, marketing, and account management. Companies are perfectly suited for the development of multidisciplinary teams, exactly because of their employees' diverse educational backgrounds.

T-profile

The third aspect to consider in assembling design thinking teams are so-called T-profiles. The T-profile symbolizes a person's ability to connect deep-rooted, analytical knowledge (the vertical part of the letter T) with broader and more integrated knowledge and abilities. "T-shaped people" are able to not only draw from knowledge in their field of expertise, but can also think and establish connections in other expert domains.

With attention paid to these three perspectives, exceptional teams can be built not only in companies, but also in educational arenas.

What are the characteristics that define good teams? McGregor (2006) puts it like this: "You enjoy being around the people, you look forward to all meetings, you learn new things, you laugh more, you find yourself putting the team's assignments ahead of other work, and you feel a real sense of progress and accomplishment." Our gut feeling often tells us whether a team has potential to be good or not. In support of McGregor, our experience shows that the following characteristics of people and teams are critical (McGregor, 2006):

Openness

Team members are open to new ideas, suggestions and solutions, even if they're "wild" and sometimes perhaps unrealistic. The same is true for openness regarding the feelings of others. Openness almost always reflects how willing a team is to listen — to one another and also to third parties.

Experimentation

The team has developed a culture of experimentation, of trying things out and allowing for failure, instead of immediately dismissing new ideas and approaches.

Help

Team members assist one another to develop characteristics such as curiosity, openness and a readiness to experiment. This often requires time and patience.

Individuality

Allowing for individuality and differences of opinion is important. The silence of harmonies between team members does not increase the quality of results. Rather, positive outcomes are fostered by constant, constructive discussion, the exploration of divergent opinions, and the acceptance of personal individuality.

Criticism

Honest criticism belongs in a good team! Without criticism that is both earnest and constructive, the danger arises that the team will not progress with enough strength and speed.

Be childlike

Teams must have the courage to occasionally be childlike. A childlike curiosity generates valuable new approaches to topics, even if they may initially appear unconventional.

Design thinking coaches and experienced psychologists can help to awaken and strengthen these properties in teams. The whole design thinking process is useless if an equal amount of attention is not paid to the composition and support of teams.

“Everything you need to know about a ‘good team’, you learned on TV”
(Schar, 2010)

Design Thinking Team 2013/2014:
The University of St. Gallen, Karlsruher Institute for Technology and University of Zurich at Stanford University in Palo Alto (2013)

Home
GEEK

IMPLEMENTATION IN COMPANIES

The introduction of design thinking into companies and organizations is already familiar territory. Over the past 10 years, numerous firms have put design thinking to use. In this time, they have garnered important learning experiences and mastered the typical challenges faced in transformational processes.

One fundamental realization is that the introduction of design thinking as a method and a culture takes much time. Changing how managers and employees think does not happen overnight—nor within weeks, or months. More often, organizations must plan for a time frame of years and allocate constantly active management resources in order to transformationally introduce the design thinking approach.

In our experience, this transformation happens on two levels — the operational, day-to-day level of the project, and the strategic level of the organization. Both intertwine when it comes to the adoption of the design thinking method.

The operational level of the project's daily routine

On the operational level of the project, there are often three variations to consider on how design thinking can be utilized:

Design thinking as a toolbox in the innovation process: In this type of application, isolated methodological elements and tools are used in existing projects. On the one hand, this approach allows for the rapid utilization of design thinking's promising elements; and on the other, employees begin to learn by making their first contact with components of the design thinking method.

Design thinking as part of the innovation process: In this variation, the diverging stage is generally combined with previously adopted development methods, such as SCRUM (www.scrum.org). The objective is supporting and pre-emptively amending the predominantly convergent development methods through a customer-oriented, creativity-fostering process.

Design thinking as a "playground" for innovation and customer orientation: In this mode, design thinking is completely deployed and aligned with a practice-oriented approach in both micro-cycles and macro-processes.

The strategic level of transformation

The strategic transformation of the company into a culture integrated with design thinking goes through four stages of development:

Awareness

Organizations in the awareness stage want to take a first step to perceptually establish the themes of customer orientation, creativity and an iterative approach. Using measures such as "town hall" style meetings with employees, targeted training by design thinking practitioners, or situation-specific projects supported by design thinking methods, employees are shown the utility of design thinking.

Experimental

At this stage of development, most companies evaluate the usefulness of design thinking in projects to solve selected problems and address certain tasks. In a continuation of the awareness stage, additional employees are trained in design thinking in order to broaden the base of methodological knowledge within the organization. Moreover, further projects are complemented by tools and process components from design thinking.

Catch-up

In the catch-up mode, companies perceive design thinking to be an established and successful method. The goal here is to stabilize and broaden its use within the organization. Connected with this is the establishment of spaces in which design thinking projects can be scaled up, on the one hand, and, on the other, the establishment of measurement systems that increase the transparency and traceability of success.

Design thinking as an organizational culture

For design thinking to become part of a company's culture, its principles need to be engaged throughout the entire organization. In addition to operational use in all forms, design thinking often drives and supports organizational transformation. This is called transformation by design. Design thinking enables internal and external problems and challenges to be processed and solved in tandem.

In addition to these developmental paths, there are numerous additional challenges to be mastered and success factors to bear in mind. These will be further explored in focused chapters:

- The organization's orientation towards customers and users;
- Middle and senior management must have support available;
- Lighthouse projects are needed to quickly show the success of the method;
- The perception and understanding of the method must be improved through training;
- Design thinking is not the method for all problems and tasks;
- Persistence is needed to implement the methods;
- Silo structures must be overcome and broken down;
- Small teams provide for impact, dynamism and the greatest success;
- Design thinking teams should have rooms at their disposal.

2

TOOLKIT

The design thinking toolkit enables us to solve design thinking problems in a structured, systematic way. Beginners can use these techniques as a guideline.

SETUP
PROJECT PLANNING

Project planning for a design thinking project orientates itself using a classical approach. This includes:

- establishing a project outline, or;
- setting the parameters for the order of events;
- generating a schedule;
- setting a project deadline;
- estimating the effort involved;
- establishing financial and risk management plans;
- determining staffing needs.

This chapter concentrates on the first three points of project planning, since their content varies the most from that of traditional projects. Additionally, we provide further advice on financial resource planning.

We can generally separate design thinking projects into two main groups (see table, p. 70):

Visionary projects
In visionary projects we work on problem statements for which the solutions lie very far in the future (five years). These projects have the objective of establishing a prognosis for products, services, and business models through the exploration of user needs and market trends. Often, the results of such projects provide the initial impulse for strategizing in an organization or sector.

Development projects
In development projects, we focus on concrete problem statements and task requirements with attention paid to users and other stakeholders. The goal is to find solutions that can be implemented in a clearly established time frame (2–5 years).

Depending on the kind of project in question, project planning is carried out differently.

The project outline
The project outline is usually succinct and addresses the following:

Existing situation: A short description of the existing situation in the organization and/or market environment, in addition to relevant framing conditions.

Problem statement: A clear description of the actual problem or challenge to be addressed.

Questions: A list of questions that must be answered to the client upon completion of the project.

Solution requirements and limitations: Requirements for output that are already established at the outset of the assignment. These can be economic, regulatory, or of a technical nature.

In an ideal case, the initial project outline should remain valid during the entire project lifecycle. However, within innovation projects, it cannot be guaranteed that new knowledge that fully invalidates the original problem statement is created. In such cases, the project outline and definition of task requirements are usually altered or added to the "problem definition and redefinition" phase of the micro-cycle. The allowance for agility and flexibility enables projects to be adjusted according to newly identified customer and user requirements the team uncovers during the need-finding phase.

Scheduling on the basis of the macro-cycle
The typical design thinking schedule for a development project is based on macro-cycle phases. This schedule has proved itself over the course of numerous successfully completed projects, and has the flexibility to be adapted to organizational conditions and requirements:

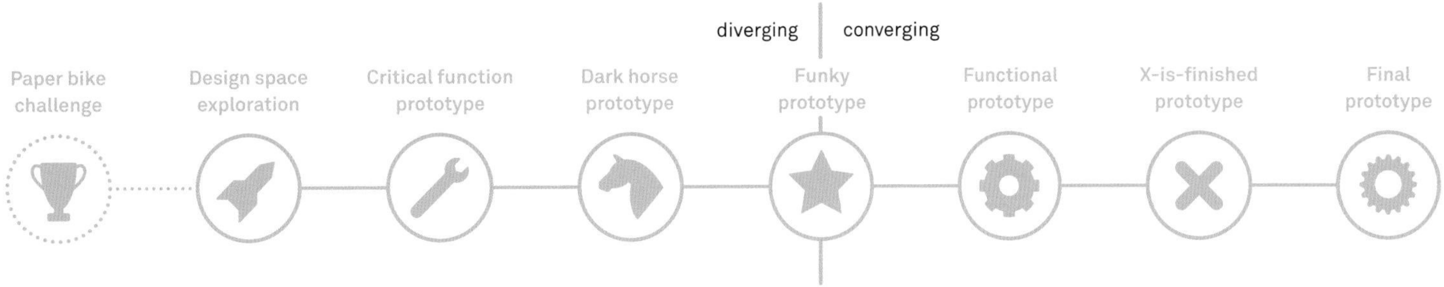

diverging | converging

Paper bike challenge — Design space exploration — Critical function prototype — Dark horse prototype — Funky prototype — Functional prototype — X-is-finished prototype — Final prototype

Milestone	Description	Output
Project start	Official start of the project — in which project planning is initiated	Project planning
Project kick-off	Get-together of the whole team in order to explain the project outline and to give the "go" signal	An understanding of roles, tasks, timing and results
Design space exploration	Exploration and definition of the problem and solution spaces with a strong emphasis on the need finding phase	Customer and stakeholder interviews, first insights, definition of critical functions, benchmarks and business model analyses
Critical function prototype (CFP)	Continuation of need finding and generation of simple prototypes that allow the most critical and important functions to be experienced and tested	Need finding results, CFPs (>20 variations), insightful failures and failure situations
Dark horse prototype	Redefinition of the initially articulated framing conditions, with realization via simple prototypes	Need finding, dark horse prototypes (>20 variations), insightful failures and failure situations

Milestone	Description	Output
Funky prototype (also called integrated prototype)	Integration of the most successful and interesting prototypes from the previous phases into more complex prototypes with the goal of identifying new implementation directions	Funky prototypes (5–10), elaborate test results
Functional prototype	Realization of a comprehensive prototype that possesses all the important functions — to test with users	Functional prototype (max. three)
X-is-finished prototype	One feature of the future final prototype must be made completely functional, so that the effort required for development of the final prototype can be better estimated	X-is-finished prototype
Final prototype	Complete, fully elaborated prototype. This shows the most important functions as realistically as possible, and provides answers to the questions posed in the project outline	Final prototype, with documentation of relevant project results

This schedule ensures that prototypes are continuously developed throughout the whole project, and that these are put to use in the collection and evaluation of feedback. This ensures, among other things, that the most important, foundational principles of design thinking are achieved: customer orientation, the iterative approach, and prototyping.

Within each aforementioned milestone, the micro-cycle (problem definition and redefinition, need finding and synthesis, ideation, prototyping and testing) is carried out at least once, but typically multiple times.

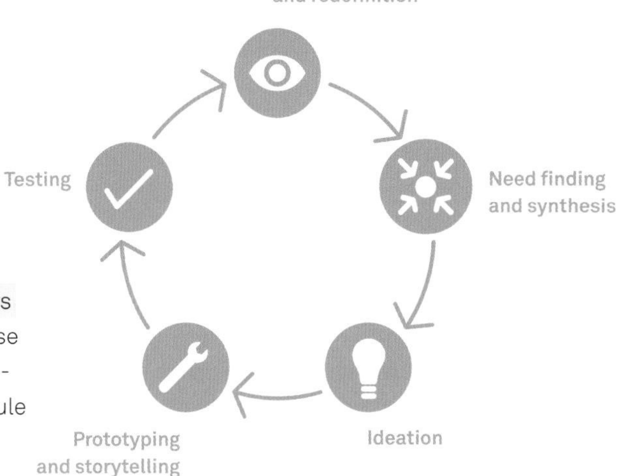

Scheduling on the basis of the micro-cycle

A second variation of the design thinking schedule is based upon the micro- as opposed to the macro-cycle. This is a particularly useful approach in situations in which access to users is limited by timing or physical circumstances. Given that there is only one micro-cycle phase in which intensive contact with users and other stakeholders is necessary (need finding), interviews and field observations are easier to plan and coordinate in comparison to a schedule based on the macro-cycle. A schedule based on the micro-cycle has the following defining characteristics:

Milestone	Description	Output
Project start	Official start of the project in planning	Project planning
Project kick-off	Get-together of the whole team in order to explain the project outline and the give the "go" signal	An understanding of roles, tasks, timing and results
Need finding and synthesis	Ethnographic field research and extraction of directions (underlying learning for further project process and development)	Interviews, results of observations, benchmarks from other industries
Ideation	Generation of numerous ideas based upon learning from the previous synthesis phase	A prioritized collection of ideas that can be further elaborated with the help of prototypes
Prototyping	Solidification of ideas into the form of prototypes in order to enable testing with users	Prototypes (generally few; up to five)
Testing	Testing of prototypes with users and further stakeholders	Test results

Besides easier planning of the contact phase with users and stakeholders, an advantage of this approach is that an in-depth focus can be placed on need finding and the associated synthesis. This makes scheduling based upon the micro-cycle a particularly good approach for visionary projects.

Timing

Naturally, the timing of a design thinking project depends on numerous factors both within and outside each particular organization. However, the experience we have collected over the course of more than 60 design thinking projects has resulted in some timing guidelines for the different phases. These can prove useful in initial planning when it comes to best estimating the time projects require.

Milestone on the basis of the macro-cycle	Minimum time needed in organizational context
Project start	1 month
Project kick-off	½ day
Design space exploration	2 weeks
Critical function prototype	2 weeks
Dark horse prototype	2 weeks
Funky prototype (integrated prototype)	1 week
Functional prototype	3 weeks
X-is-finished prototype	3 weeks
Final prototype	3 weeks

The timing table above assumes the following parameters:
Team size: 4–5 people
Average time commitment per person: two days per week; sometimes higher intensity in organizational contexts
Interviews: The time frame assumes that conversations and observations with users and additional stakeholders can be organized and carried out without obstacles. The chapter on need finding and synthesis more fully addresses the subject of interview methods.

Financial resources

The financial requirements of design thinking projects are generally no different than those of other organizational projects carried out locally and/or globally. However, two budgetary items are typically planned separately, since they can lead to increased costs:

Travel budget: Customer orientation means that the design thinking team has to be physically present at the location of the customer or user. This cannot be delegated to market research companies or others! Therefore, some projects involve higher costs due to the travel required. Realistically, the team should plan a few visits. Only this level of engagement can lead the design thinking team to the required level of empathy for the user and a deep understanding of the problem statement. Sometimes it is critical that the team work on location, be it in Ethiopia, Bhutan, Colombia or South Africa.

Prototyping material: Building prototypes is a central activity in most design thinking projects. While in early phases of the project the prototypes are generally inexpensive and easy to build, the costs increase in later phases. This applies to both products and services. The costs of materials, and external services required in some cases (e.g. production of software or video), should be included early in cost calculations.

TOOLS

- Planning software, such as Microsoft *Excel* or *Project*, is helpful
- *Post-its* can be used to effectively display project planning. For example, activities can be written on *Post-its* and then arranged chronologically on a foam board. The advantage is that everyone in the room can see the project plan, making it accessible and easy for the team to work on.
- SCRUM (www.scrum.org)/Kanban boards, for planning project tasks (either on a whiteboard or with Trello.com)

PROCEDURE AND TIPS

- Start with project planning as early as possible. Give yourself at least one month.
- Establish a list with the most important contact people.
- Develop a project outline with an order of events. This does not need to be complete from the very beginning, but it does need to provide a basis for conversation with the clients hiring your design thinking team.
- Estimate the time, costs and effort associated with the project.
- Confirm important appointments, such as deadlines and milestones, with the client employing your team.
- Establish a schedule and deadline that take human resources into consideration, so that the project´s objective can be achieved in the given time frame.
- Produce a budget.

This is often an iterative process, and multiple realizations are necessary before the project can begin.

Dimensions	Visionary project	Operational project
Goal	Basis for a customer-oriented strategy for products, services and business models with inspiration and ideas from the design thinking process	Solution of contemporary challenges in B2C and B2B contexts via generation of customer-oriented products, services and business models
Target time frame for solution	> 5 years (up to 10 years)	1–5 years
Process configuration	The emphasis is on need finding, synthesis and ideation phases; prototyping is less central	The focus is on all phases of the design thinking cycle
Challenges	The relation to the organization and assignment of tasks must be sustained; abstract ideas should be made experienceable through prototyping	In the diverging phase, clients and stakeholders can identify and comment on what they perceive as an incorrect or missing focus; the design process can then be viewed as a mere implementation process for existing ideas

SETUP

INFRASTRUCTURE

Carrying out a successful project requires appropriate infrastructure. Particularly in organizations with no suitable spaces for design thinking projects, it can be a challenge to establish the right environment. Nevertheless, the appropriate environment for a successful collaboration can be set up with simple materials.

Private work space for each design thinking team member: A private workspace should be available for every member of the design thinking team. Despite intensive collaboration within the team, it must be ensured that team members can also work in peace and concentrate on the numerous tasks required of them.

Workspace for collaboration: In addition to private workspaces, it is important to build the right environment for teamwork and collaboration. In this field, team members often have to work with and around one another, for example, when discussing interview results and arriving at the conclusions that guide the project´s next steps.

Foamboards: Results, ideas, items for discussion and so forth must be documented and kept available to the team at all times. In addition to flipcharts and whiteboards, we strongly recommend foamboards. They are light and flexible, and the team can both cover them with *Post-its* and quickly repurpose them.

Creativity kit: Numerous tests and workshops are carried out outside the team's own spaces, so it is helpful to compile the most important materials into a suitcase of supplies to bring along.

Storage options: The amount of space required for equipment (foamboards) and prototypes should not be underestimated. Many office spaces are not set up to accommodate the amount of material required, acquired and generated during design thinking projects, so these issues need to be considered during planning.

Materials (e.g. for prototyping): Basic materials such as *Post-its*, paper, and so forth should be available in every office. Numerous other design thinking authors request a crafting or creative area; a hobbyist's bench, or craft store. Having access to lots of materials is certainly good, but we have learned that a small, limited assortment is sufficient, and these limits can push the team's creativity a step further.

Prototyping workspace: Particularly with products, the team needs a workspace in which to construct prototypes. This can create challenges. For example, if electronics such as *Arduino* are used, then a soldering iron needs to be quickly put into use, without violating fire safety regulations.

Telephone and conference calls: Today, it almost already sounds old-fashioned to list this as a requirement. However, numerous projects have shown that simple means of communication such as video conferencing facilities are not always available to design thinking teams. In many cases they are, however, indispensable for interviews with customers and users, as well as experts.

Customizing the workspace: A small budget should be available to enable the team to acquire things needed to give their rooms personality and to create a closer workspace.

TIPS AND TRICKS

- Review the materials and workspaces available — in person if possible. Our experience has shown that there is often a substantial discrepancy between the materials that room booking services list, and what is actually present in rooms.
- Make a list of materials to order and infrastructure needed, and request them. Make sure that you take into consideration the fact that certain materials are simply not in the standard supply closet of many organizations — for example, foamboards. In such cases, be prepared for substantial wait times of up to a month.

NOTES

SETUP

TEAM SETUP

The team takes on a special meaning in design thinking projects. When a team is put together well, it can always achieve more than what individuals alone are capable of. But what does 'well' mean when it comes to building teams? Over the past few years, we have used the "Teamology" indicators, based on Carl Gustav Jung's personality typologies, as a guide for putting teams together. Teamology determines an individual's personality type, and roughly classifies them upon the basis of extroversion versus introversion, sensing versus intuition, thinking versus feeling, and judging versus perceiving. For us, the personality types of individuals are the decisive criteria when it comes to building teams. We know that some other authors consider multidisciplinarity more important than personality type, however our experience has shown that a lack of multidisciplinarity is easier to compensate for in our global world (e.g. by integrating experts in certain fields), than the personality composition within the team itself.

Besides this, experience has proven the value of the tried-and-true approach of holding a meeting on the subject of team dynamics. In this group meeting, the standard rules of communication in teams are revisited for emphasis and discussion. Additionally, the team discusses the results of the personality tests — with natural attention to confidentiality. The goal is for the team to equip itself to identify and to compensate for its potential weak points, and better manage potential conflicts that may arise.

Personality type

The Teamology test can determine an individual's personality type. This can change according to life stage and environment, but it serves as a good indicator of a person's basic disposition. The test classifies people in the following fields:

Extrovert versus introvert: Extroverts are more open, generally enjoy contact with others more, and can be more adept at communication than introverts. On the other hand, introverts often have strong analytical abilities.

Sensing versus intuition: Intuitive people make decisions with their "sixth sense" — their gut feeling. This typically comes from experience and self-confidence. More fact-oriented, sensing people often delay making decisions. Their gut feeling does not immediately arise, and typically comes only after they study the facts of a situation.

Thinking versus feeling: Thinkers generally make decisions rationally, based upon the facts presented to them. People who feel more strongly tend to also draw the emotional aspects of the situation and of their colleagues into decision-making.

Judging versus perceiving: Evaluative people tend to make quick judgments and hold onto them over long periods of time, even in light of new facts. People in the "perceiving" group tend to absorb situations over long periods of time before making decisions.

"Adding a few people who know less, but have diverse skills, actually improves the group's performance."

(Surowiecki, 2004)

Over the past several years, the categories of Teamology tests have been given characterizing names as "mastermind", "teacher" or "supervisor". What is important to note is that all categories are regarded positively. When setting up a design thinking team, as many quadrants of the teamology test should be covered as possible. Quadrants or fields that are not covered should be discussed during a team dynamics meeting.

In organizational contexts, personality tests have to be carried out carefully — above all in regard to protecting individuals' data. In most cases, human resources departments as well as employee associations need to be involved, even when the individual employees themselves freely agree to the tests.

Multidisciplinarity

The second criteria for the composition of a design thinking team is multidisciplinarity, by which we mean the subject-related education and expertise of the team members. Relevant to a project's context, a range of qualifications and profiles have to be combined — for example, by putting a software engineer, a designer, an ethnographer and an economist into the same team. Behind this approach is the hypothesis that the team should have access to knowledge from multiple disciplines.

Practically, this means that various profiles of qualifications will help to build the team that works on the problem. Unfortunately, in many industries, such as banking, a starkly one-sided approach has negatively impacted the diversity of qualification profiles available.

T-profile

The third criteria is related to one's ability to bring more than just expertise or "deep knowledge" to his or her team (as symbolized by the vertical part of the "T"). To a certain extent, team members should also be able to think and work within other fields (as symbolized by the horizontal part of the "T"). "T-shaped" individuals are usually difficult to identify from afar. When composing a new team, some indications exist in CVs, or the project initiator might know an individual's characteristics from previous projects.

Sources

McGregor, D., "The Human Side of Enterprise", 2006 (annotated ed.)
Parker, G.M., "Team Players and Teamwork: New Strategies for Developing Successful Collaboration", 2008 (2nd ed.)
Surowiecki, J., "The Wisdom of Crowds", 2004
Wilde, D.J., "Teamology: The Construction and Organization of Effective Teams", 2007

TOOLS

- Calculation table for evaluation of test results (Wilde, D. J., 2007)

TIPS AND TRICKS

- The use of personality tests requires rigorous attention to confidentiality with all data collected!
- Check with your client organization's management whether or not personality tests are allowed — and under what circumstances. In some cases, also check with employee representatives.
- If your client organization allows personality tests, before testing you should still have a personal conversation with each employee to fully explain the test and interpretation of the results. Should an individual have concerns, try to alleviate them. If their concerns cannot be resolved, do not do the test — listen to your gut feeling.
- Carry out the test.
- Analyse the data. The results form the foundation of the team's composition.
- Discuss the test results individually with each participant and the team results with the group.

NOTES

1

2

3

4

5

6

7

8

---

1 Kick-off with team members from China, Finland, Switzerland and Germany at Stanford University in Palo Alto
2 Team meeting in Finland
3 Shanghai, after a long day of workshops and interviews
4 Traveling between Zurich and St. Gallen after need finding
5 Meeting with company representatives in Switzerland
6 Workshop in Espoo, Finland
7 One of many virtual group meetings
8 Meeting up for work in Zurich

9

10

11

12

13

14

15

16

9 Workshop in Espoo, Finland
10 Trial for fall presentations in Zurich
11 Meeting up in San Francisco
12 Rehearsing a midterm presentation in St. Gallen
13 Short tour through the University of Zurich before the next group meeting
14 Need finding in Shanghai, China
15 Meeting up in San Francisco
16 All teams' final presentation at Stanford University in Palo Alto

SETUP

POST-ITS

Setting up a design thinking project requires the right approach to working with some key tools. This includes *Post-its*, which a number of producers now offer. Unfortunately, use of these colored sticky notes increasingly polarizes users — some feel that while *Post-its* make work fun, this does not necessarily correlate with productivity. In design thinking projects, *Post-its* play an important role and their advantages are immediately at hand. The decisive point is whether or not we use *Post-its* correctly.

Advantages

Overview: When stuck onto foamboards or walls in a logical order, *Post-its* efficiently generate a quick overview. Since they make content physically available in the room, as opposed to hidden within a computer, team members can often more quickly familiarize themselves with new results or changes.

Easy restructuring: Post-its can be easily moved around and restructured, for example when clustering learning. In contrast to *Post-its*, which can easily be re-sorted, recomposed in new orders and "re-stuck", flipcharts and whiteboards often fall flat, as content is written on a blank surface and repeatedly has to be generated anew.

Transportability: Post-its are exceptionally easy to transport. For example, if a workshop needs to be carried out externally, existing *Post-its* can simply be brought along from the project workspace.

Mobility: Teams that always carry *Post-its* in their luggage are able to quickly and easily jot down learning while in transit. These can be integrated into the project workspace later.

What should we pay attention to?

Clarity: Those who use small or messy handwriting when using *Post-its* run the risk that their teammates will not be able to decipher them. This prevents teammates from making use of their learning and ideas.

Sketches: When possible, central learning should be sketched on *Post-its*. Sketches of mechanisms and graphs, for example, generally help other team members to quickly gain an understanding of subjects. They also enable the team to more easily return to earlier themes.

Content: A well-structured *Post-it* generally has a headline. The headline is followed by a half-sentence that quickly and concisely lays out essential ideas.

Marker: This seems like a non-essential detail, however experience has shown that the use of overly thin or fat markers tends to make *Post-its* illegible. While ballpoint pens almost always lead to *Post-its* that are illegible from a short distance; fat-tipped markers result in *Post-its* that are illegible from any distance. Medium-sized markers, such as the Edding 1300, have proved themselves. The color of the marker is also important to consider — a red Edding on a red *Post-it* will be difficult to read.

Handwriting: Handwriting on a *Post-it* should be clearly legible. In most cases, this means using block letters. This rule of thumb holds true: Even an extremely good idea is worthless if no one can read it!

Dies ist eine passende Stiftstärke für gute Lesbarkeit

undeutliche Handschrift kann man nicht lesen....

Überschrift

könnt ihr lesen, was ich hier schreibe? Ich glaube nicht. Es ist einfach zu klein oder was denkt ihr?

Bitte deutlich schreiben

Überschrift

Erkläre kurz die Idee mit einem prägnanten (Halb-)Satz

ÜBERSCHRIFT

... aber Beschreibungen können auch so lange sein, dass niemand die Zeit hat, sie zu lesen – komm auf den Punkt!

TIPS AND TRICKS

- The *Post-it* rules should ideally be clearly communicated and illustrated during the kick-off meeting, for example while doing physical warm-up excercises such as the "Stickman Warm-up".

TOOLS

- Markers (for example the Edding 1300)
- *Post-its* in a variety of sizes
- Surfaces suitable for putting up *Post-its* (these can be hard to find in some organizations). Foamboards provide an ideal solution.

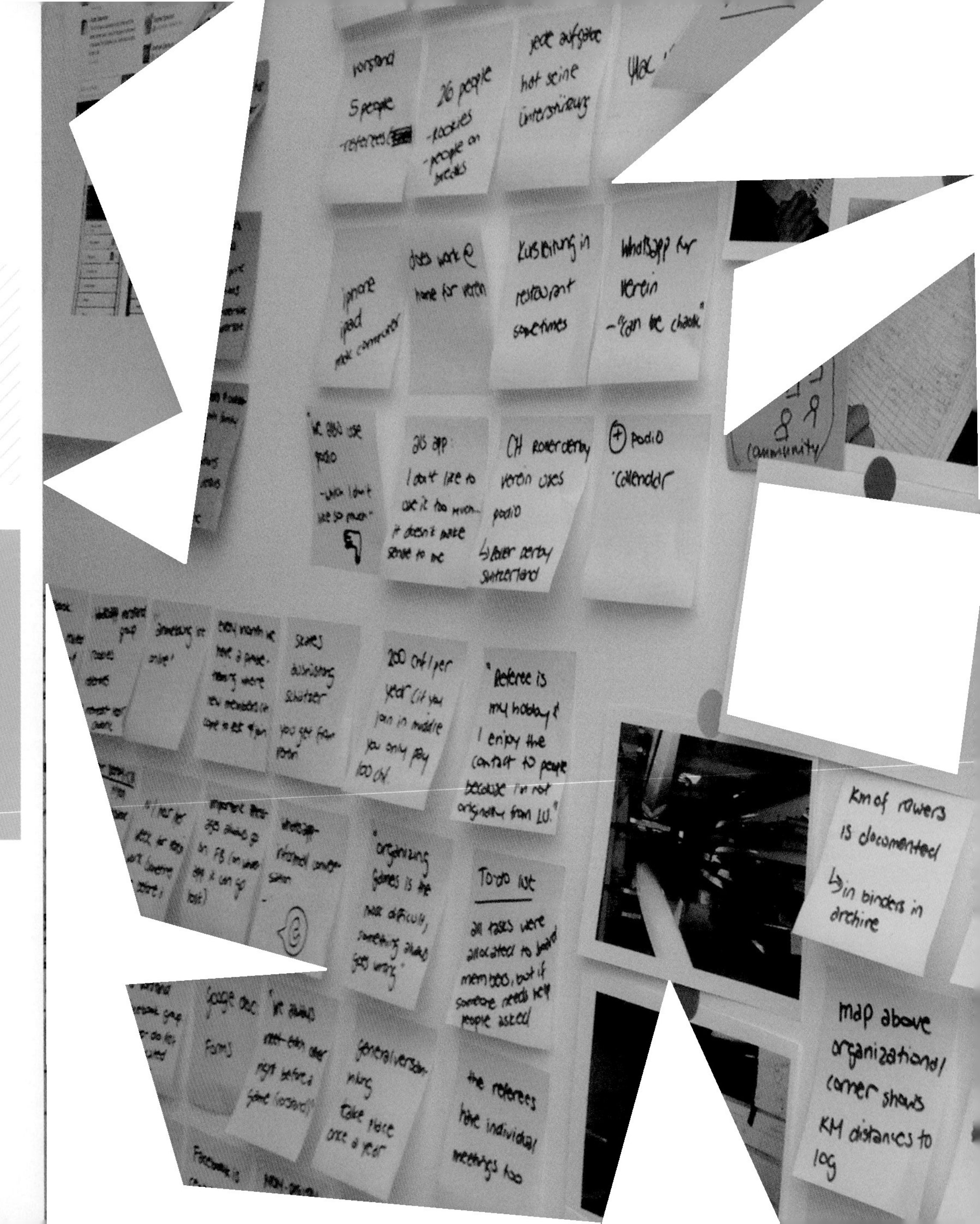

"if we have to discuss things we have informal meetings over breakfast after trainings or mails circling around"
"meetings are not very scheduled or planned. there are no hierarchies"
50 active members, ~50 who are non active
archive located in a small cabinet next to lockers.
· binders
· unorganized
we are most proud of our boats
Some boat spots are hired by other rowers or clubs
"we have an organization corner"

SETUP
DNA ANALYSIS METHOD

How did this method emerge?
With enthusiasm, productive creative processes are very personally felt and guided.

Often, teams miss an initial collective understanding of what is to be achieved, what fits the organization, and what parameters should be considered as a given.

As a consequence, the process of assessing approaches and ideas is often inefficient and disappointing.

To prevent these almost standard situations from arising, several years ago milani design & consulting AG in Zurich started to develop a method: the DNA analysis method.

What is the essence of the method?
The DNA analysis method makes it possible for team members to rationally and emotionally experience organizational goals within a four-hour workshop. The tone is creativity and consensus-oriented, and guarantees content homogeneity and high motivation in teams.

The rational and emotional targets and results are comprehensibly documented and serve as an efficient evaluation of all subsequent approaches. Personally motivated, off-topic discussions are prevented and replaced by an effective, constructive process.

Key points of the method:
- Developed over 14 years
- Consequential connection of rational and emotional factors, including traceable quantitative measurability
- Continuous development and improvement
- Documented history of successful use
- Implemented and realized in innovation projects, product development, graphical user interface (GUI) design, communication, corporate identity/corporate design, corporate fashion, exhibitions and corporate architecture

What will it achieve?
- Increased decision-making speed and confidence via collective understanding of what fits the organization, for example in relation to innovation management
- Fast and effective optimization of the organization's direction, for example, in relation to crises or generational change
- Strengthened motivation and identification
- Definition of quick measures and adoption of action plans
- A knowledge of the organizational DNA makes it possible to confidently achieve positive differentiation from competition

What's it about?
- Concentration on the essential — the DNA of the organization
- Often, the organization's personality is understood for the first time; the management team experiences how their organization's personality "feels" to others, and how this can be used and/or changed

The process:
Step 1/Analysis of all strategy papers:
It is essential to filter out relevant information: What moves the organization? What are the recurring themes?
For example, in relation to findings on innovation: Is the organization past- or future-oriented, risk-averse or adventurous, agile and flexible or static, and so forth.

Step 2/Visual presentation of the strategic content:
Thematic pairings of strategy content, such as "risk-averse versus adventurous" are visualized in metaphors/images on an evaluation axis.

Step 3a/Four-hour workshop with leadership team:
As a first step, management makes a critical self-evaluation of each thematic pairing (such as "risk-averse versus adventurous"). They subsequently review the results in relation to the actual base characteristics of the complete contemporary organization:

- Open, honest discussions, consensus, and definition of the organization's current strategic position establishes the "is" situation. With knowledge of the organization's current elemental character, or DNA, a second step reviews its specific characteristics, such as its innovation portfolio (how innovative are our products?) in relation to the "is" situation. Another aspect could be the innovation effect of the organization's general image (in terms of products, communication, exhibitions, architecture and so forth). This process makes the points of convergence and deltas (gaps) impressively visible — and hence the need to make it authentic through actions. This means bringing actual organizational potential to the fore, as well as future capabilities.

Step 3b/Comparison to competitors:
With the newly-learned "strategic glasses", competitors are evaluated using the same criteria. There is a focus upon the recognition of overlaps and potential areas for differentiation.

Step 3c/Establishment of the "should" position:
Based on the deltas between the essential organizational character and specific characteristics and a market comparison, "is" and "should" positions are defined and the related steps to close the gaps are set out.

Step 4/Documenting the process and results:
All content should be protocoled and presented in an overview. Visual representations that express the measurable rational, relative and emotional components will be helpful. If possible, the agreed target positions should be not only expressed in text and visual representations, but also with evocative images that effectively communicate the emotional aspects.

Experiences with the method

The DNA analysis method uncovers important discrepancies between strategic positioning and organizational identity on the one side, and such issues as the pre-existing innovation culture or impact of market presence and existing products on the other.

The deltas and the related need for action become quantifiable and measurable, and are then transferred into specific operational targets.

The management of values and any change in values must be broadly anchored in order to integrate new orientations into the organization without using exclusively top-down mechanisms.

Therefore, we recommend repeating the DNA workshop in subsequent levels of the hierarchy while consistently managing the targets in all organizational areas, following the principle of "train the trainer".

The result: Certainty that targets of consequence are dispersed throughout the organization and that above all, innovation processes, such as those based on design thinking, can be understood and adopted.

Testimonials

"The DNA analysis method has helped us enormously in terms of aligning values across the Medela Group. This has led to increased efficiency and effectiveness through the subsequent development of a global, visual model of our market-facing image."
Renate Schreiber, CEO, Medela AG

"This was the best process with externals that I can remember".
Heino von Prondzynski, VR Roche and CEO a.D., Roche Diagnostics Ltd.,

"At Feller, we carried out a very important phase of organizational positioning and alignment using the DNA analysis method. The simple and efficient process was only exceeded by the clarity of the results. We could glean valuable learning about the "is" versus the "should" state of the organization... and through this, we achieved a collective understanding within the Feller management team."
Roger Karner, CEO, Feller AG

Which themes are relevant in my organization? What is our DNA based on? Where are internal DNA, processes and solutions out of alignment?

NOTES

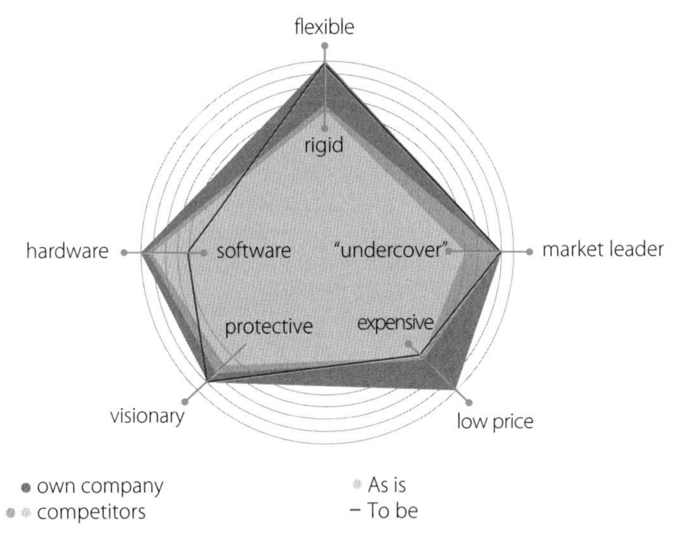

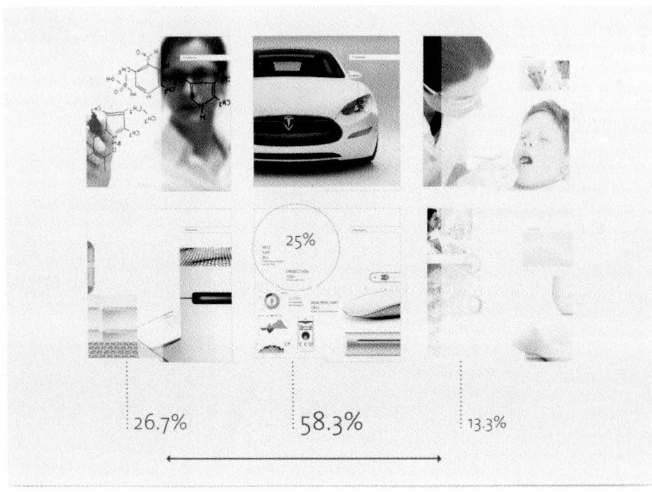

Example of a company's DNA analysis

PROBLEM DEFINITION AND REDEFINITION

Every successful design thinking project begins with a good, goal-oriented question statement. However, the initial question statement provides only the starting point for a design thinking project. Through multiple iterations of the micro-cycle, the question is constantly adjusted, or further refined through sub-questions. The definition and redefinition of the problem statement, through continuous adaptation and refinement, is a central step in the project.

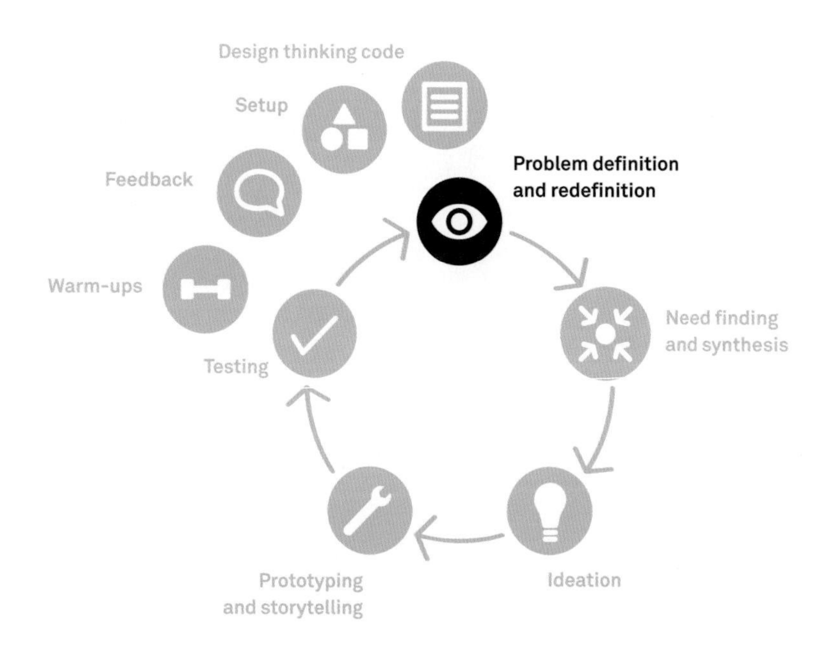

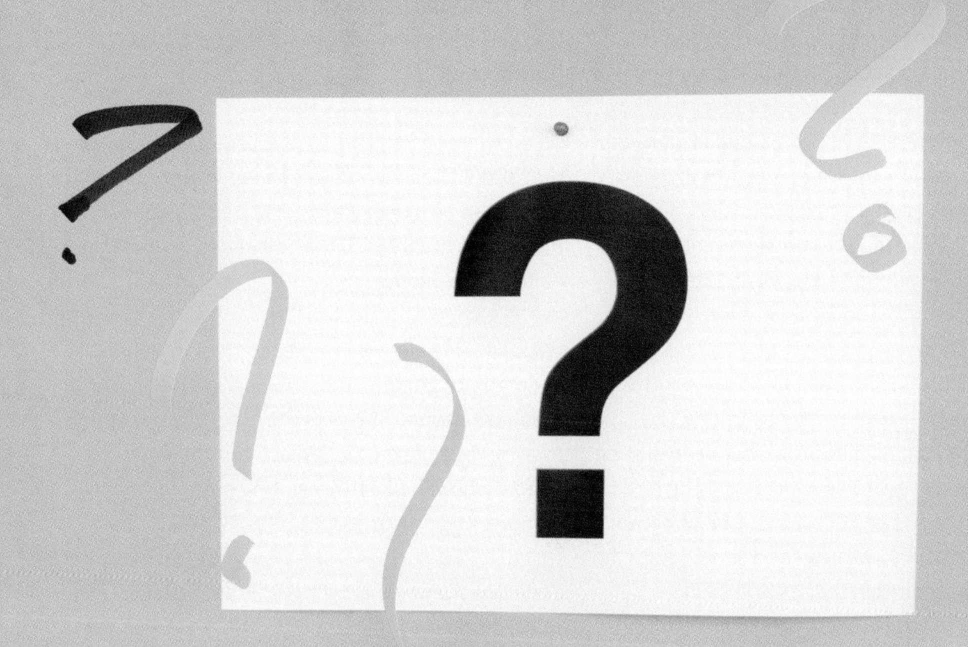

PROBLEM DEFINITION AND REDEFINITION

DEFINING THE QUESTION STATEMENT

Defining and redefining the initial question is a central element of design thinking. If questions are too focused on a technological solution or a certain search area, this can restrict the design thinking team early in the process. It quickly becomes hard for the team to think outside the proverbial box. Conversely, questions that are too broadly defined around abstract goals are just as problematic. To prevent teams failing at the outset, we offer the following guiding principles on effectively structuring and logically limiting the solution space:

Do not offer solutions

The question statement should not propose a solution. Neither technical nor economic solutions should be part of the question that frames the design thinking process. It is the design thinking team's job to search for and locate solutions. During this process, there is a danger that client propositions are misunderstood and even used in place of a successful question framing strategy for the project. It is generally far easier and more effective for the design thinking team to develop new ideas without being offered a possible solution.

Inspiration and emotion

If a question statement is inspiring, it will not only appeal to team members pragmatically, but also emotionally. Exactly this kind of emotional connection can ignite the additional motivation required to intensively work with and explore the subject matter over the long term. At the same time, emotions open up new dimensions for consideration. Furthermore, the question statement should be clearly defined and formulated.

Give a direction

At first glance, this rule appears to run contrary to the first rule — "do not offer solutions." However, there's a big difference between stating a solution and suggesting a direction! The direction of a question can, for example, be directed by an organization's strategy and show where the organization sees its products and services heading in the future. There is no concrete solution connected with direction, however there is a clear orientation towards a specific area in which a solution might be found.

Developing guidelines

The figure on p. 89 shows a structured template for developing a design thinking question statement. The template should be considered a guideline, and has led to the successful development of question statements in numerous workshops. Furthermore, every question begins with a "how might we..." statement. Leading with these words often helps to set the right tone for the question.

Object/Something: This always refers to an object that should be newly developed or invented. The object can be a product, a service or a business model. The qualities of the object can be provided to offer direction.

Persona: This element relates to the target user or customer group — those for whom the project result is relevant. The persona is not the client that commissioned the project, but the person or people who will use the resulting output upon completion of the project.

Conditions/problems/changes: Framing conditions, problems and so on that influence the central object of interest. In the banking sector, these might be regulatory requirements. In the chemical industry, it might be environmental conditions that impact the scope of work.

When filling the template, one should not be too focused on each individual element. However, numerous successful projects have shown that this template is helpful for developing and discussing new question statements.

"Good ideas originate from good questions."

Company Name
How might we re-design
something
for
persona
considering
conditions / problems / changes / setting / ...

TIPS AND TRICKS

- Discuss possible question statements for the project with your client.
- Use the template to structure the question statement, either during or after the meeting with your client.
- Repeatedly re-examine your question statement with your client and other stakeholders in the organization, and keep revising it.
- Always evaluate your question statement in light of the criteria for a "good" question statement.

Defining question statements is an art! Don't get frustrated, even if it seems frustrating and challenging at first. Our experience has shown that putting effort into finding the right question statement tends to be rewarded with good project results.

PROBLEM DEFINITION AND REDEFINITION

STRETCH GOALS

Using the "stretch goals" technique, existing boundaries are symbolically extended in vertical and horizontal directions.

Vertical extension: Vertical extensions always refer to products, services or processes that already exist — and sets new targets for them. For example, an organization's customer satisfaction rating could be raised from 90 to 100 percent. A further example might be the reduction of airplane fuel use from 0.7 gallons per passenger per 60 flight miles to 0.2 gallons per passenger per 60 flight miles. The targets set must be both technically and organizationally challenging to achieve at the time.

Horizontal extension: This concerns the development of completely new products, services or processes. In addition, new target markets or framing conditions can be established in order to enliven the ideation process. An example from the finance sector would be offering an asset management service portfolio to classic retail banking clients.

What both extension directions have in common is the redefinition of goals to foster new problem-solving approaches.

Some risks are associated with this method:

Demotivation: Targets that are unrealistic or uninspiring can quickly demotivate a team.

Stress: Particularly highly motivated team members, committed to reaching "hard" targets, can become overly stressed when trying to solve extremely difficult problem statements. This can adversely affect their performance levels.

In practice, these methods have proved themselves as excellent reframing techniques.

Sources

Kerr, S. and Landauer, S., "Using Stretch Goals to Promote Organizational Effectiveness and Personal Growth", 2008

TIPS AND TRICKS

- The team writes the question statement on a large sheet of paper or a whiteboard.
- For vertical extension, list the goals to achieve one under the other.
- Think within your group about how goals can be clearly and ambitiously extended. Write this next to the original goals.
- Define your new question statement together.
- In practice, a proven method for the stretch goals exercise is to invite members of other teams to participate. The fresh and sometimes naïve perspective helps set targets anew.

Indian Institute of Technology: Understanding and defining the problem in the context of an international project (2015)

PROBLEM DEFINITION AND REDEFINITION

FRAMING AND REFRAMING

In problem framing and reframing, the principal concern is to achieve a new perspective on the question statement. By considering the initial problem in a different light, new directions for exploration and fresh ideas emerge. These can lead to original solutions in new conceptual spaces.

The reframing process is carried out in three steps (see Morgan, G., 1997):

Identifying assumptions: In the first step, explicit and implicit assumptions and framing conditions for the project are identified. These include the assumptions and conditions of the initial question statement, as well as every assumption that has come to light over the course of the project. These assumptions may have been previously unknown, or only superficially considered.

Questioning/challenging assumptions: In the second step, the assumptions collected from the team are collectively challenged with the goal of identifying and understanding the reasoning behind each assumption. It is not critical to find one single truth; instead the team should achieve a basic understanding of the formation process behind the assumptions.

One technique that can be used when challenging assumptions is the 5W, or five times why method.

3. Redefining assumptions: In the third and final step, the assumptions deemed interesting and important are morphed. The exact form of this new definition depends on the phase in the macro-cycle. For example, in the dark horse prototype phase, explicit assumptions are inverted to express the opposite. Why? It is often easier to think about the opposite of something than to think about it in the actual intended direction. For example, many teams aim to improve situations. This goal is, of course, valid, but the power of imagination is quickly used up. When this happens, it often helps to consider the exact opposite goal. How can the situation be made worse? From these answers, the team can glean inferences about the original positive case. Reframing helps teams to think outside the box, however team members must be prepared to leave their comfort zones.

Sources

Morgan, G., "Images of Organization", 1997 (2nd ed.)

TIPS AND TRICKS

- Write the initial question statement on a whiteboard.
- In your group, collect the assumptions that have been accepted as fact up until this point, as well as any assumptions which were explicitly given in the project outline.
- Within your team, prioritize the assumptions in relation to their importance and subjective relevance.
- Select the three most highly prioritized assumptions and challenge each one with "why?" up to five times. Write the why questions and answers next to each other.
- Stop the 5W method when the results become too general and offer little new knowledge.
- Within your team, select the three most interesting why answers.
- Formulate the three most interesting why answers into new question statements (using the "How might we..." technique).

INITIAL QUESTION STATEMENT

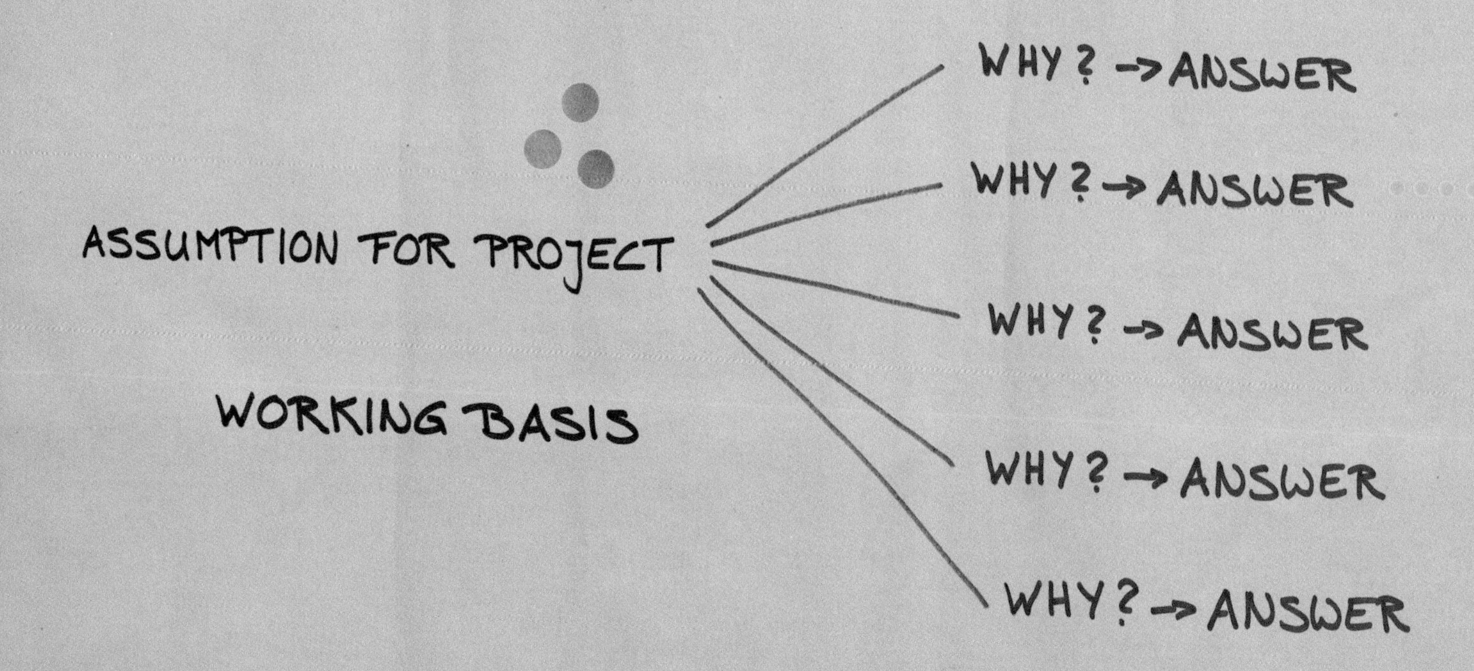

PROBLEM DEFINITION AND REDEFINITION

GET INSPIRATION FROM THE FUTURE

When trying to put yourself mentally into the far future, it can be helpful to watch a science fiction film or read a science fiction novel together as a team. If one considers the retrospective film *Preview of the World Tomorrow* (Kerstan & Schmuckler, 1972), the power of imagination that helped some forward thinkers from the past envision the today's world becomes apparent. Many compelling films and books fulfill this criteria.

From today's perspective, depending on the question, we recommend films such as:

- *Space Patrol — The Fantastic Adventures of the Starship Orion*
- *2001: A Space Odyssey*
- James Bond (all)

Films and books offer excellent fodder for a design thinking team trying to establish ambitious new targets and framing conditions for their project definition.

Sources

Direction 2000 — A Preview of the World Tomorrow (*Richtung 2000 — Vorschau auf die Welt von morgen*) (documentary, 1972, Kersten, P. & Schmuckler, A.)

2001: A Space Odyssey (1968, Stanley Kubrick)

TIPS AND TRICKS

- Popcorn, chips and chocolate are a good start for a movie night! Make sure you have enough.
- A projector and a large screen should be available.
- Foster a movie theater atmosphere and watch the films together.

NEED FINDING AND SYNTHESIS

Need finding ensures customer orientation. The three techniques of observation, interviews and engagement systematically help us to comprehensively identify users' needs. Synthesis condenses the data collected into learning, and offers direction for the phase of idea finding.

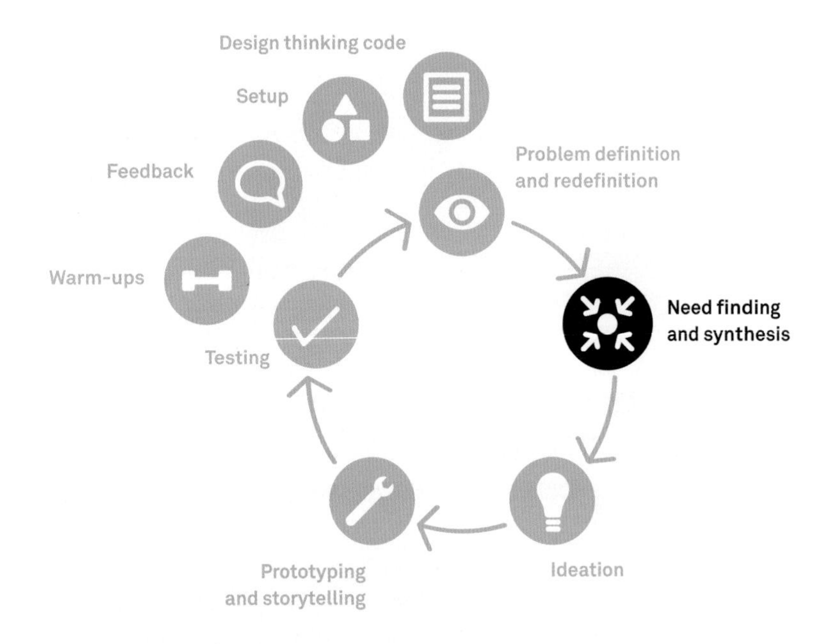

Customer No. 1
ATM

NEED FINDING AND SYNTHESIS

NEED FINDING CYCLE

The process underlying need finding, the need finding cycle, consists of the following four steps, which ensure a solid approach.

1. *Scope and prepare the exploration space:* Establish the user and stakeholder groups to be examined and the approach to be used. The following research questions are in the foreground: Who should be observed or interviewed? In which environment should this investigation take place? Which sampling methods should be used? How big will the sample of each user or stakeholder group be?

2. *Observation:* Photos, videos and notes should be generated. In sensitive fields, what's observed can also be sketched.

3. *Interviews and role changes:* Carrying out interviews based on lists of questions. Deeper understanding through role change within the design thinking team: the team takes on the role of the interviewee on the issues under investigation. For example, team members could travel with a service technician for a day, putting themselves in his or her shoes.

4. *Interpreting the results and redefining the exploration space/synthesis:* Sharing observations and interview results within the team. Making sense of the results, and learning and drawing from them as a team. This step is called "synthesis" in some procedural models, and is frequently facilitated by structured frameworks.

Need finding is a fundamentally iterative process in a design thinking project. That means that with every step of the micro-cycle, there are cases in which need finding goals may need to be adjusted or corrected. This flexibility can result in expansion or contraction consequences for the need finding strategy. In practice, and with growing experience, the design thinking team can dynamically implement the need finding process.

It is important that the research question for need finding covers the following criteria:

- Which objects (item, service, etc.) will be introduced and used in the area under consideration?
- Which environmental conditions influence the behavior of the people observed?
- How do people interact with one another in the environment under investigation?
- Which types of person can, in some cases, be classified or differentiated?

Sources

Brown, S., Gray, D. & Macanufo, J., "Gamestorming: A Playbook for Innovators, Rulebreakers, and Changemakers", 2010
d.school Stanford University, "What to Do in Need Finding", 2015

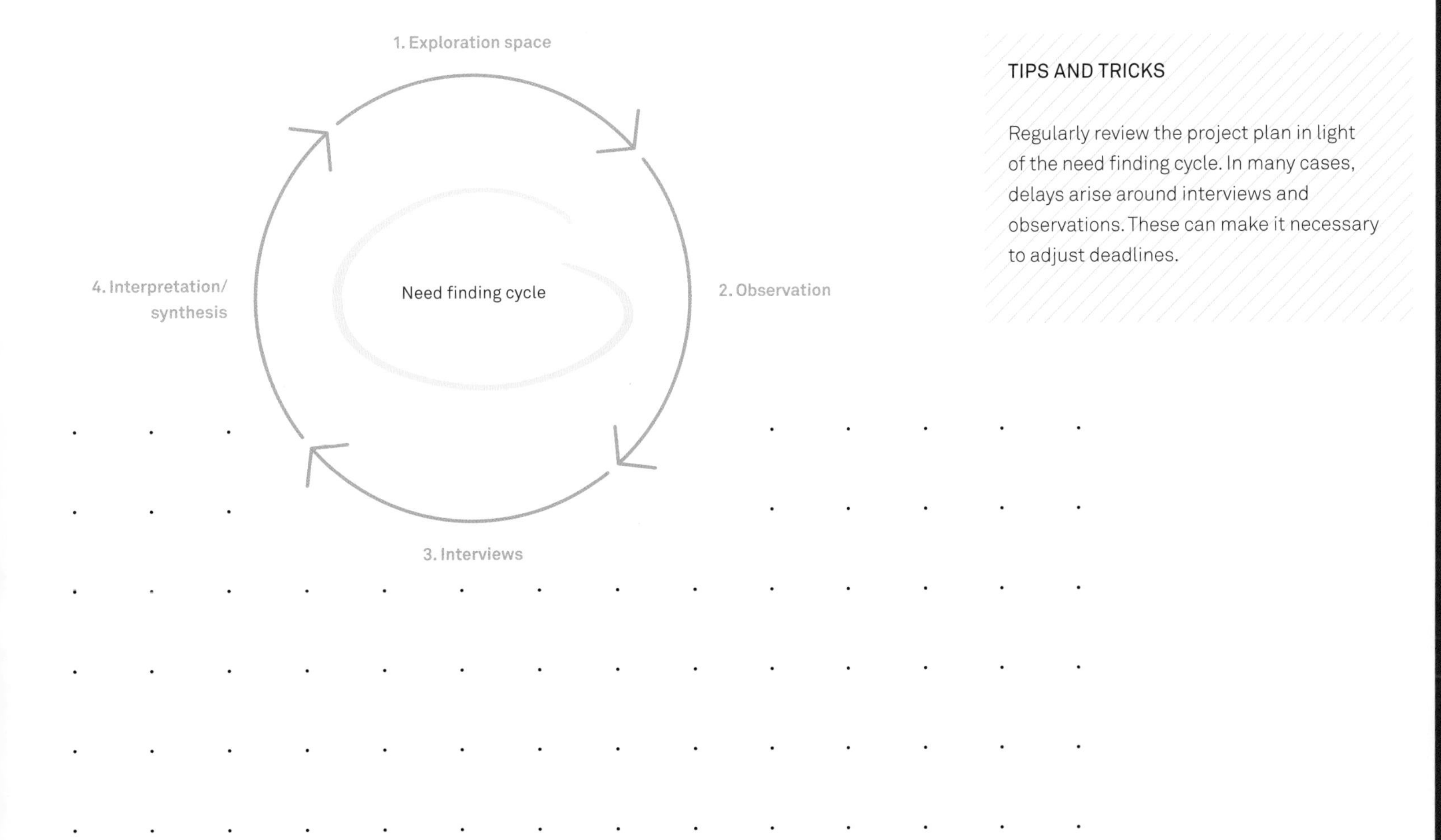

TIPS AND TRICKS

Regularly review the project plan in light of the need finding cycle. In many cases, delays arise around interviews and observations. These can make it necessary to adjust deadlines.

GUIDELINES FOR EXPRESSING NEEDS

After need finding, the gathered needs are put down in writing. At this time, the needs should be discussed without drawing any connection to potential approaches to addressing them.

In line with the work of Ulrich and Eppinger, a few guidelines have proven helpful in practice :

"What" — not "how": A description of a need should not include or specify a solution. Therefore, all descriptions that imply a "how" should be avoided.

Specificity: The description of the need should be as specific as possible. It is best to use concrete examples from the contexts of interview partners. General, paraphrased descriptions should be avoided at this stage.

Positive, not negative: The expression of needs should be positive — formulated as a constructive contribution.

Product attributes: When possible and when it makes sense, concrete attributes of products or services that could be improved or integrated should be referred to. Here, great care must be taken not to prematurely close the solution space.

Avoidance of "must" and "should": Avoid words like "must" or "should". The word "should" relativizes needs, and this doesn't make sense in the need finding context. A need is either relevant or irrelevant. In some cases, using the word "must" to express needs overly limits the solution space for potential solutions.

Use of these guidelines is not mandatory, but they can help to effectively express the user and stakeholder needs identified.

Sources

Ulrich, K.T. & Eppinger, S.D., "Product Design and Development", 1995, 4th ed.

TIPS AND TRICKS

- Particularly for beginners, it makes sense to repeat the guidelines at both the start of the project and at the beginning of the need finding phase.
- "Learning by doing" with real data produces the strongest learning effect on these rules.

NEED FINDING AND SYNTHESIS

SAMPLING TECHNIQUES

Interviews can be carried out either one on one or as discussions in larger focus groups. Sampling determines the interview partner. There are several different sampling methods.

Purposive sampling

In purposive sampling, people or groups are selected through predefined criteria such as age, gender or lifestyle interest. The size of the selection depends on time and number of people available. A common criterion for sample size is Ulrich and Eppinger's recommendation of a minimum 8–9 one-on-one interviews per cluster. The number of participants can be smaller in focus groups.

Quota sampling

In quota sampling, people are selected through predetermined criteria, just as in purposive sampling. However, the number of people per cluster and the issues to explore are also determined in advance.

Snowball sampling

Snowball sampling uses the power of social networks. Snowball sampling expects initial participants to either suggest additional interview partners, or the same marketing measures that uncover initial participants will also reach similar individuals for further conversations. This sampling method is used when searching for "hidden" groups — groups with whom contact would otherwise be difficult to establish (such as individuals from underground subcultures).

Sources

Ulrich, K.T. & Eppinger, S.D., "Product Design and Development", 1995 (4th ed.)
Guest, G., Mack, N., MacQueen, K., Maney, E. & Woodsong, C., "Qualitative Research Methods: A Data Collector's Field Guide", 2005

NEED FINDING AND SYNTHESIS

TARGET GROUP IDENTIFICATION

The project definition provides an initial target group. As additional knowledge is developed over the course of the project, this target group can be extended or changed. While the design thinking team works exploratively in early phases of a project and seeks out conversations with multiple users and stakeholders, in later phases the exploration space becomes more clearly structured and tailored to the frame established during need finding.

A quick and generally valid model for subdividing the exploration space is the diffusion of innovations model DOIM; (Rogers, 2003). The DOIM classifies groups of people in a particular context into the following five categories:

Innovators: These are people that are interested in trying out and purchasing the newest technologies and services. Often, high social status and financial resources enable this experimentalism. Innovators usually have strong contact with scientists and other innovators.

Early adopters: Early adopters are similar to innovators in that they have high social status and financial resources, however they are more selective in choosing new products and services. Early adopters are the opinion leaders for the subsequent groups in the model.

Early majority: The early majority require substantially longer to adopt new products and services than the aforementioned groups. Their educational level is generally above average.

Late majority: This group adopts innovations substantially later than the average of the wider population. Adoption is often accompanied by substantial skepticism. This group generally has below average financial resources.

Laggards: People in this group often demonstrate no pronounced opinion leadership and are the last to adopt new products and services — if they do so at all. Most people in this group tend to have very low incomes.

In practice, the three middle groups comprise 90 percent of the total population, with innovators and laggards comprising 5 percent each. Particularly in the early phases of exploration, the design thinking team can draw upon such a distribution of user groups. The need finding strategy is to identify and interview as many people from all groups as possible. At a minimum, people from the innovator, laggard and one of the middle groups need to be examined.

Research shows that 8–9 interviews per group generally suffices for identifying 80 percent of needs (see Ulrich & Eppinger, 2008). This means, for example, that in a typical project with five groups, 45 to 50 conversations should be held.

To better structure the (potential) target groups, they can be distributed within a "customer selection matrix". This visibly presents the criteria encountered, and the frequency in which they arise. This enables the team to achieve a manageable overview of the massive number of interviews and observations, and preserve their learning. Furthermore, the progress made during need finding is easily visualized.

Sources

Rogers, E.M., "Diffusion of Innovations", 2003, (5th ed.)

Ulrich, K.T. and Eppinger, S.D., "Product Design and Development", 1995, (4th ed.)

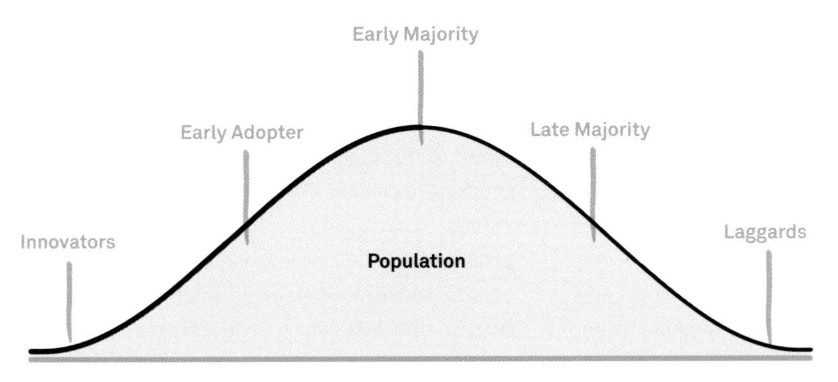

Diffusion of Innovations Model based on Rogers (2003)

TIPS AND TRICKS

- The team can ideate on possible identification criteria for a target group. In some cases, standard models such as the diffusion of innovations model can be used as a guiding framework.
- Based upon the ideation, the team can compile ideas on how the target group might be approached or communicated with. Perhaps the team members know innovators or experts in a certain field. Then of course, the approach is particularly easy.

FOCUS GROUPS

Focus groups are already widely used in industry. In design thinking, explorative focus groups are used with the goal of identifying the needs of as many users and stakeholders as possible within a short amount of time. This leads to valuable learning about the sample group.

When putting together a focus group, there are a number of possibilities:

Traditional focus group: 10–12 users are brought together for a tightly scheduled workshop intended to achieve complete results. The average length for a traditional focus group is approximately two hours. These focus groups are especially appropriate for identifying needs, but are also useful for generating new ideas. These groups are also successful in sensitive environments (such as in medical questions).

Mini focus groups: The number of users is generally six to eight people, over a one- to two-hour time frame. When used as a complement to traditional focus groups, mini focus groups allow a higher level of detail in discussion due to the lower number of participants, and more specific questions.

Dyads: Dyads are also frequently called "friendship focus groups". They are primarily used for analyzing problems in which lone participants have issues articulating the answers. Often, this approach is used with children or teenagers. These conversations are usually directly honest and full of learning.

Super groups: Super groups (over 100 participants) are predominantly used in testing. Due to the size of the group, question sets must be well-structured. These groups should be accompanied by at least two moderators. In terms of results, these groups quickly deliver goal-oriented feedback, however content cannot be explored in depth.

Triads: With triads, three people who are similar or different on previously determined criteria are invited to participate in a focus group. Triads are similar to one-on-one interviews but deliver a broader range of information. In the case that the participants differ on focal criteria, their answers can be directly compared. If participants are similar, singular contributions generally build upon each other and deliver high-density content in a short amount of time.

Party groups: With party groups, a participant invites the rest of the group (including the moderator) to a meal or event at his or her home. During this party, the issue in focus is discussed.

"The purpose of focus groups is not to infer but to understand, not to generalize but to determine a range, not to make statements about the population but to provide insights about how people perceive a situation."

(Krueger & Casey, 2000)

These conversations are above all concerned with achieving constructive, critical and rational inquiry. However, party groups are generally only appropriate for consumer-related topics and question statements.

Online discussion groups: Another variation is discussions led online. In principle, most of the options previously described can also be carried out online.

In addition to the focus group variations described here (explanatory and explorative), there are additional activities that can be carried out in focus groups:

Feature prioritization: Determining the characteristics of products or services that are especially relevant in the perspective of the customer or user.

Competition analysis: Analysis of the characteristics that users especially value or avoid in products or services of competitors.

Trend analysis: In newly established trends, focus groups can help to elicit and understand underlying reasons.

Sources

Brenda, L. (ed.)., "Design Research: Methods and Perspectives", 2003
Krueger, R.A. & Casey, M.A. "Focus groups: A Practical Guide for Applied Research", 2000

TOOLS

- Audio recording device
- Camera
- Notebook

TIPS AND TRICKS

- Plan focus groups early. The organizational effort is usually greater than expected. Establish a set date.
- Write to the potential participants to inform them of the goal and your confidentiality procedures
- Organize the space in which the focus group will be carried out early.
- Continually follow the status of sign-ups. If not enough participants sign up, in some cases additional ones will need to be located and contacted.
- Inform the team running the focus group about the goals and agenda early, so that everyone knows what needs to be done on the day of the event.
- Think about the organization of participant reception upon arrival. This is especially relevant for customer and user focus groups!
- Lead the focus group through the workshop.
- After you conclude a focus group, it is advisable to write participants a thank-you note.

NEED FINDING AND SYNTHESIS

INTERVIEWS

The goal of interviews is to gain an understanding of the motives, behavior, intentions and opinions of individuals within the background of their given context. Because interview research is an established method in the social sciences, there is extensive literature available on the preparation, implementation and processing of interviews.

In design thinking projects, interviews are carried out with representatives of all relevant groups. This includes users and experts, but also people that generally avoid products and services new and unfamiliar to them. "Open" questions are used most. In contrast to closed questions, open questions give interview participants no concrete options to answer with. A simple rule is: just let the interviewee talk. Often, the most interesting details are hidden within the interviewee's account. These often only come to light after recapitulation and extensive retrospective consideration of the interview.

When it comes to conducting interviews, we point to a model by Michael Barry (adapted by the d.school at Stanford University, 2015b). Barry structures the interview "flow" in six phases. These both manage and stimulate the conversational flow:

1. *Introduction:* Short interviewer introduction and synopsis of the issues of interest for the interview. For sensitive and/or confidential topics, the protection of the individual and their data is again ensured.

2. *Kick-off:* Pose a relevant "icebreaker" question. This often involves the interviewee's concrete work or life context.

3. *Rapport:* The interviewer next prompts the interviewee to describe a specific event, happening, or a recent use (of a product or service) relevant to the general area in question.

4. *Grand tour:* In this phase, after the interviewee is thematically and emotionally familiarized with the question, the interviewer tries to address issues focused around distinctive characteristics. For example, he or she might ask, "Could you describe the distinctive function?" Or "What makes the function so unique?" Immediately after the "rapport" phase, many interviewees have established sufficient associations with the field in focus to enable discussion of emotional aspects of the product or service during the grand tour.

5. *Reflection:* During reflection, the interviewer tries to animate the interviewee to reflect on his or her account. Example questions focus on themes such as, "What would you improve?" or, "How would you completely redesign this product or service?"

6. *Wrap-up:* The wrap-up phase concludes the interview and typically reveals what will be done with the results of the conversation. Particularly in personalized interviews (in which the name of the interviewee is known), a transcript is sent to the interviewee in order to ensure that the interviewer understood correctly what the participant meant.

In addition to the method with which the interview is carried out, conversations can take different directions:

Retrospective interview: These have the goal of elaborating and better understanding events in the recent past. This kind of interview often helps the interviewer put sequences in order, thereby enabling the reconstruction of a course of events.

Lead user interview: Here, lead individuals are interviewed. As leaders, these people are often the first to use a given technology, or to offer a reliable prognosis for the future on the basis of experience.

Camera study: To supplement an interview, a camera is given to the interviewee with the request that they take photos of specific product or service features. For example, the interviewee in a project related to nutrition could be asked to document his or her every meal. These photos are collected and then provide the basis for the interview. The advantage of this technique is that a given topic can be addressed in a very detailed way.

Expert interview: Expert interviews are carried out with experts in a given field. In addition to enabling deep discussion of the given subject matter, expert interviews give the interviewer the opportunity to quickly expand their knowledge of the issues in question.

Wake-up interview: Initial interviews allow the interviewer to gain a better understanding of the context and scope of the question statement. Typically, these are quick telephone conversations or conversations at home with a few family members. Wake-up interviews are generally used at the beginning of a project.

Sources

Brenda, L. (ed.), "Design Research: Methods and Perspectives", 2003
d.school Stanford University, 2015b

TIPS AND TRICKS

- In practice, it is generally worthwhile to carry out interviews with two interviewers. Using this approach, one-sidedness can be avoided during the interpretation of interviews.During the interview, the second interviewer can keep the interviewee's expression in sight, match this with the statements made and use this to determine which statements might lead to interesting learning.
- Furthermore, a proven approach is to use a survey to test the questions with a few interviewees at the very beginning of an interview series. If the questions are not precise enough or leave important subject areas unanswered, the survey can be adjusted as needed.

TOOLS

- Audio recording device
- Camera
- Notebook

1

2

3

1 Interview with a doctor on the use of cold medications, India (2015)
2 Interview about clubs in Zurich, Switzerland (2014)
3 Interview with a medicine man in South Africa (2014)

NEED FINDING AND SYNTHESIS

OBSERVATION

Observation aims to describe and document happenings. Human behavior is frequently at the center of the investigation. The advantage of observation is that the individuals observed are usually operating in their own environments and therefore display an authentic character. Furthermore, participants do not have to sacrifice additional time to participate in the research.

By observing people while they use products or services, learning on opportunities for improvement can take place. Challenging these situations can also, however, uncover completely new needs and lead to new knowledge on these users. In later phases, the results achieved can be carried over into new solutions or improvements to existing solutions.

A disadvantage is that this kind of research has only limited possibilities for usage in the field, the reason being legislation protecting the rights of individuals. During observation, it is therefore critical to respect the legislation within the given country, as well as any rules within an organization (see the design thinking code).

A further challenge of observation lies in the interpretation of the data collected. Usually, this data is not able to be used without further interviews. For this reason, interviews follow observation in the need finding cycle.

In practice, how can observation be carried out?

Cameras: To the extent that legal stipulations and guidelines within an organization allow, cameras can be used to permanently make a record and then analyze a situation.

Hanging out: A long stay in the environment of those under observation. Whoever has spent more than three hours in a Starbucks can understand which learning on people and their behavior can be acquired via this method.

Engagement: A tour, or following in the footsteps of the user simulates observation.

Be a tourist: In some cases, it is worth asking an insider for a "tour", in order to better understand the work processes. For example, at a restaurant or a laboratory.

Paparazzi: By this, secret or anonymous shooting of photo or video is meant. Here, it is extremely important to pay attention to legal issues; that is, no individual should be recognizable in the film or photos.

What results should observation obtain?

Environment: The physical space or place
Actors: The people involved
Activity: Activities in relation to the observed object
Object: Physical object of relevance
Event: Events in relation to activities
Time: Sequence of events and activities
Goals: The goals of the people in the observed situation — what they have achieved or tried to achieve?
Important: With all methods, it is absolutely crucial to ensure that no individual rights are violated!

Sources

d.school Software Design Experiences Stanford University (n.d.)

TOOLS
- Video camera
- Notebook

TIPS AND TRICKS

- Before you begin observation, ensure that all necessary materials and technology are complete and fully functional.
- Plan your observation techniques as well as the goals of observation. Share the task responsibilities within your team, if necessary.
- If necessary, gain approval to carry out observation.
- Be certain to inform your team of behavioral expectations in the observation environment in advance.
- Carry out the observation.
- Evaluate the results of observation.
- Remove all personal data as soon as the data has been evaluated. At the very latest, remove all personal data upon completion of the project.

NOTES

NEED FINDING AND SYNTHESIS

ENGAGEMENT

Engagement refers to the design thinking team's personal engagement with stakeholders such as users and customers. It supplements interviews as a technique by which design thinking team members can experience the situation of the user. For example, a team member may put themselves in the shoes of a service technician for household appliances, meaning he or she might carry out the same tasks and activities as the trained employee. This helps the team to achieve a deeper understanding of the work processes, behavioral tendencies or the complete context. In parallel, in this phase team members often develop a deep empathy for the people and the situations they are exploring.

In industrial environments, such engagement is often difficult or impossible for safety reasons, especially if special courses on safety or customer data confidentiality are necessary. Given that such visits are a central component of the whole process, it is important to allow enough time for the necessary preparation.

In a consumer environment, engagement is often much easier. A team member could, for example, take on the role of an interested customer in a bank and carry out the process of opening an account from start to finish.

What are we looking for in engagement?

Insight: Deep understanding of the use and interactions with products and services.

Process knowledge: Knowledge of the process from the perspective of the user or users who carry it out.

Empathy: Empathizing with the situation of the user or group of people affected.

TIPS AND TRICKS

It is important to allow plenty of time for engagement. Therefore, within your team, try to identify the situations in which you want to achieve engagement well in advance, and begin the planning process accordingly.

TOOLS

- Video camera
- Still camera
- Notebook

NEED FINDING AND SYNTHESIS

BENCHMARKING

Benchmarking means looking for comparable situations in other fields and industries, and gaining ideas, solutions or insights from them.

An example might be a project for optimizing work processes in an operating theater. Some criteria for comparison are order, sterilization, precision, and a zero-failure culture. A benchmark for optimizing processes in this environment might be a pitstop at a Formula 1 track. Within only a few seconds, the mechanics must get the race car into perfect running condition using an immense level of precision. There is no room for failure in the processes, nor in the results. By comparing these situations, we can gain valuable learning to optimize processes in the operating theater.

In order to find appropriate benchmark objects or situations, we recommend using a four-step approach. Design thinking teams continually carry out this process as routine during a project:

Search: The search is concentrated on market competitors, technologies, patents, and so forth. With the help of Google Patent Search, for example, design thinking teams can quickly and efficiently locate most patents. Blogs, web communities and newspapers also constantly publish useful information.

Evaluation: The information found is assessed on functions and relevance to the field of focus. One recurring question is: What inspires the team for their own set of tasks?

Follow: Consistently follow new developments on the internet. Web portals such as bloglovin.com make this process easier.

Share: New information from this process should be shared and discussed within the team as quickly as possible.

The focused search for appropriate benchmarks is based on the search for specific analogical criteria in other situations (such as in the operating theater/Formula 1 example above), which can shed light on the problem and its associated tasks. Using these criteria, the team can begin to collect focused benchmarks. Once again, this calls for the four-step process.

TOOLS

- www.flipboard.com
- www.feedly.com
- www.pinterest.com

The goal of benchmarking is to improve need finding, design and documentation.

Benchmark of a digitized car:
Tesla Model S (2014)

NEED FINDING AND SYNTHESIS

FRAMEWORKS

The primary goal of a framework is the consolidation of information collected in need finding. This enables the important learning to be recognized. Frameworks are generally multidimensional, and are often expressed as diagrams. They enable the grouping and organization of information collected.

For example: In an open-plan office, there are both organized and what appear to be very chaotic desks. It is also observed that both organized and disorganized people appear to work at both "types" of desks. At first glance, this may not seem to make sense. However there are people that can maintain an excellent overview of their work, even at a messy looking desk.

However, particularly in Western cultures, an orderly workspace is typically equated with an organized person. In contrast, a messy workspace is seen to represent a disorganized person. If one transfers both desk types and personalities into a diagram, or framework, and adds the existing offers of leading furniture companies, he or she will notice that there are not many solutions for organized people with chaotic workspaces who wish to maintain an orderly appearance to align with expectations.

Frameworks, such as that described in the example above, help the design thinking team to maintain an overview in complex situations. Venn diagrams, 2x2 (or higher) dimensional matrices and bar charts can all be useful when presenting information. In many cases, free-form representations can also be effective.

Sources

Beckman, S.L. & Barry, M, "Innovation as a Learning Process: Embedding Design Thinking", 2007

"The ultimate goal of the framing step is to reframe, to come up with a new story to tell about how the user might solve his or her problem or to come up with a new way of seeing the problem, which in turn will allow the team to come up with new solutions."
(Beckman & Barry, 2007)

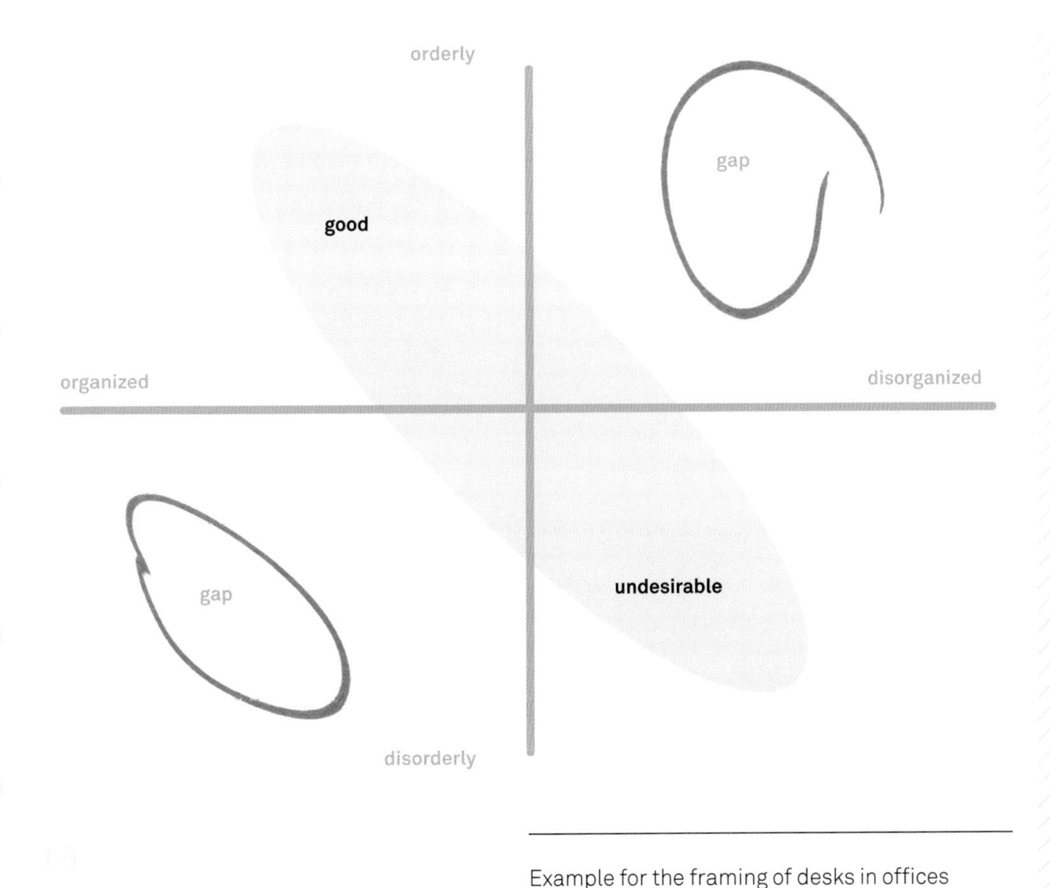

Example for the framing of desks in offices

TIPS AND TRICKS

The procedure for developing a framework is straightforward, but it works only when sufficient data is available in the form of interviews, focus groups, observations, and so forth:

- Visualizing all learning identified in the need finding process. *Post-its* are especially useful, as each piece of learning can be represented by one *Post-it*.
- Clustering results in line with various criteria. This step is teamwork and does not always lead to a useful framework on the first attempt. Ideally, the team will experiment with different dimensions and forms of representation.

Finding well-founded frameworks to depict relevant content often requires a lot of time. Do not be discouraged if you are not satisfied with the first-round results of framework identification.

NEED FINDING AND SYNTHESIS

FIELD NOTES

The documentation of impressions from observations, focus groups and interviews in the form of field notes is an integral part of need finding. In academia, ethnography is an established field which is centrally concerned with the efficient and reliable collection of field notes. There are some general tips developed in practice that help make collecting field notes easier and more effective:

First impression: The design thinking team should immediately record their first impression upon entering a situation. This can include smells, moods, behaviors, organization and arrangements.

Exactness: The team should record all impressions as exactly as possible. Whether they have to do with concrete actions or interactions, room setups or anything similar, it is important to record and retain these with as much accuracy as possible.

Do not generalize: Generalizations should be avoided in field notes. Generalizations can be made later in the design thinking process, when data is evaluated, but not first-hand in the field.

Aspects: Fieldnotes should address the same aspects as observations and interviews: objects, people, interactions and the like.

Sources

Emerson, R.M., Fretz, R.I. & Shaw, L.L., "Writing Ethnographic Fieldnotes", 1995

TIPS AND TRICKS

- Each team member should always carry writing materials (pens and a notebook).
- Notes should only include the most important words and interview results.
- We recommend writing up field notes into more thorough notes directly after a given interview/observation, and certainly on the same day.

TOOLS

- Notebook

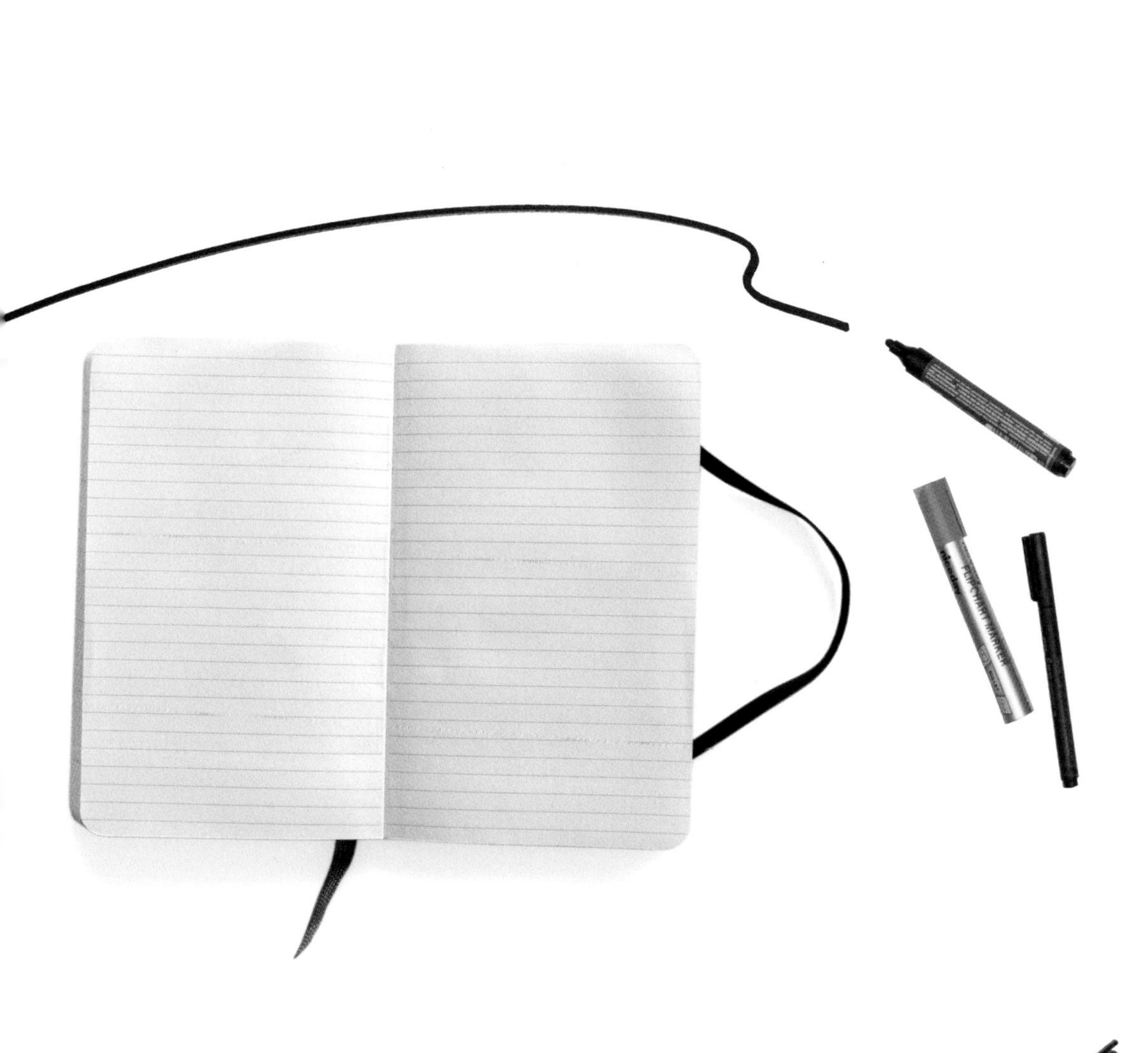

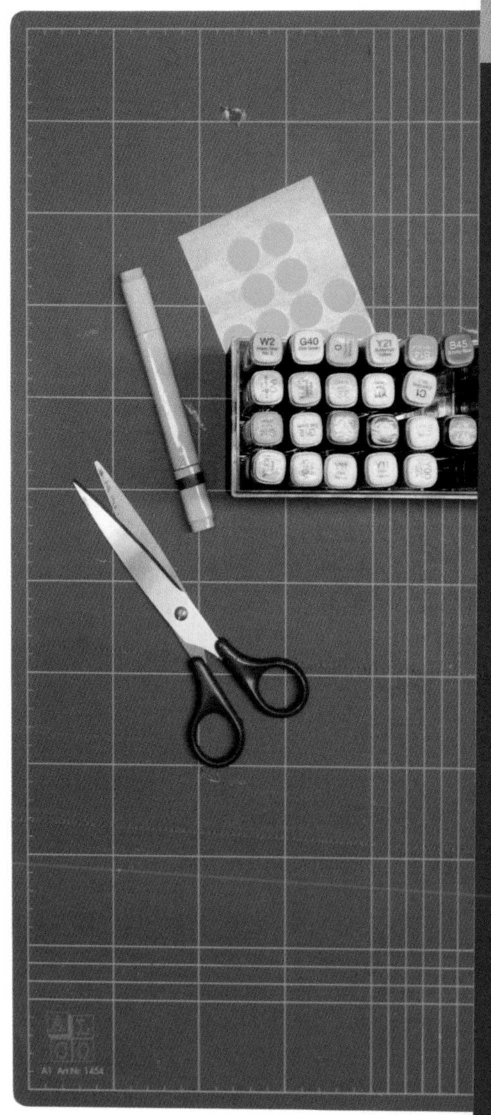
W2
G40
Y21
B45

MOODBOARD

The moodboard is a tool drawn from the communication and design sciences. The goal is to evocatively express the environment, atmosphere and mood to select user and customer groups. Similar to a large collage, pictures, videos and/or text are put together to illustrate something that is often too difficult to describe with other approaches (see Gray et al., 2010, p. 200–201). Generally, a large piece of cardboard is used. This should always be accessible to the design team during the design space exploration phase. This is so that, on one hand, it can be constantly adapted, and, on the other, to serve as a source of inspiration and reminder of goals.

In the later project phases, for example functional prototyping, the moodboard can also be updated using raw materials that hold a particular importance to the application and use of the developing product. For the automotive industry, this material might be fabric or leather for the seats of a car. Additionally, digital moodboards, such as those on pinterest.com, can be used. This is particularly appropriate for use by international teams working from different locations across the globe.

Sources

Brown, S., Gray, D. & Macanufo, J., "Gamestorming: A Playbook for Innovators, Rulebreakers, and Changemakers", 2010

TOOLS

- www.pinterest.com
- Moodboard (iPhone application, www.atinytribe.com)

TIPS AND TRICKS

- Together with your team, select the theme areas you want your moodboard to display.
- Pin a theme at the top of a foam board or large piece of paper.
- As a group, look for pictures, texts and other materials that illustrate the chosen theme, and pin them to the board.

Moodboards typically grow iteratively over a long time horizon.

MATERIALS

- Cardboard and foam boards
- Newspapers
- Pictures and other inspiring material

123
O2

NEED FINDING AND SYNTHESIS

AEIOU

AEIOU is a technique that is particularly helpful for beginners to observation. The term stands for Action, Environment, Interaction, Object and User.

The human brain focuses and concentrates on only a few of what it perceives to be essential aspects of a situation, in order to make it easier to comprehend. However, this mechanism of excess information filtering is a handicap to observation. Observation is about gaining a complete overview, and not being limited by each particular observer's subjectivity.

To achieve this, the AEIOU technique distributes different tasks among different observers. These observers concentrate exclusively on the aspects of a situation assigned to them.

What does AEIOU mean?

Action: Observation and recording of individuals' activities.

Environment: Observation and analysis of the environment. In a hospital, this might be the filing system for handling materials in closets and storage spaces.

Interaction: Observation of the interactions between people and objects.

Object: Observation of the use of objects, such as machines.

User: Observation of the user in his or her operating context.

TIPS AND TRICKS

- Use the AEIOU framework to divide the group into small sub-groups.
- Each sub-group observes and analyzes the material assigned to them.
- Afterwards, the results are compiled during a discussion within the entire group.

NESPRESSO
&
APPLE
EXPERIENCE
Couples
VALSER

NEED FINDING AND SYNTHESIS

EMPATHY MAP

Empathy is the ability to perceive and understand the feelings, emotions, thoughts, and thereby also the behavior, of others. Many techniques help us examine individuals at a technical level, but the empathy map allows us to also understand people and groups at an emotional level.

In the approach from Brown et. al (2010) demonstrated here, an empathy map draws on four central experiential components of individuals and groups (or personas) — seeing, hearing, thinking and feeling — in relation to a specific context. For this form of mapping, what the person or group being examined sees, hears, thinks and feels is recorded in either a physical or virtual diagram. Later, what individuals actually saw, heard, thought and felt during interviews and observations is also recorded. Thereby, the design thinking team gains a deep understanding of the emotional dimensions of the question being explored.

Seeing: What does the person see, right before his or her eyes?

Hearing: What does the person hear?

Thinking: What does the person think about their acoustic and visual perceptions, within the context of their experiences and expectations?

Feeling: How does the person feel in relation to what they see, hear, and think? Do discrepancies result in some cases? Do these perhaps lead to satisfaction or dissatisfaction?

Sources

Brown, S., Gray, D. & Macanufo, J., "Gamestorming: A Playbook for Innovators, Rulebreakers, and Changemakers", 2010 O'Reilly, d.school Stanford University, 2015b

TIPS AND TRICKS

- Bring sufficiently large writing and drawing materials, such as a flipchart or piece of cardboard, and draw a large cross from the upper left to the lower right corner, and the lower left to upper right corner.
- Label the four resulting fields Seeing, Hearing, Thinking and Feeling.
- Pin the name of the person or persona (group) to be analyzed in the middle of the cross.
- Within the group, analyze the four categories. Each team member also has *Post-its* in hand, and adds his or her contributions to the four fields as they arise.

TOOLS

- *Post-its*
- Flipchart marker
- Flipchart or other material to draw on

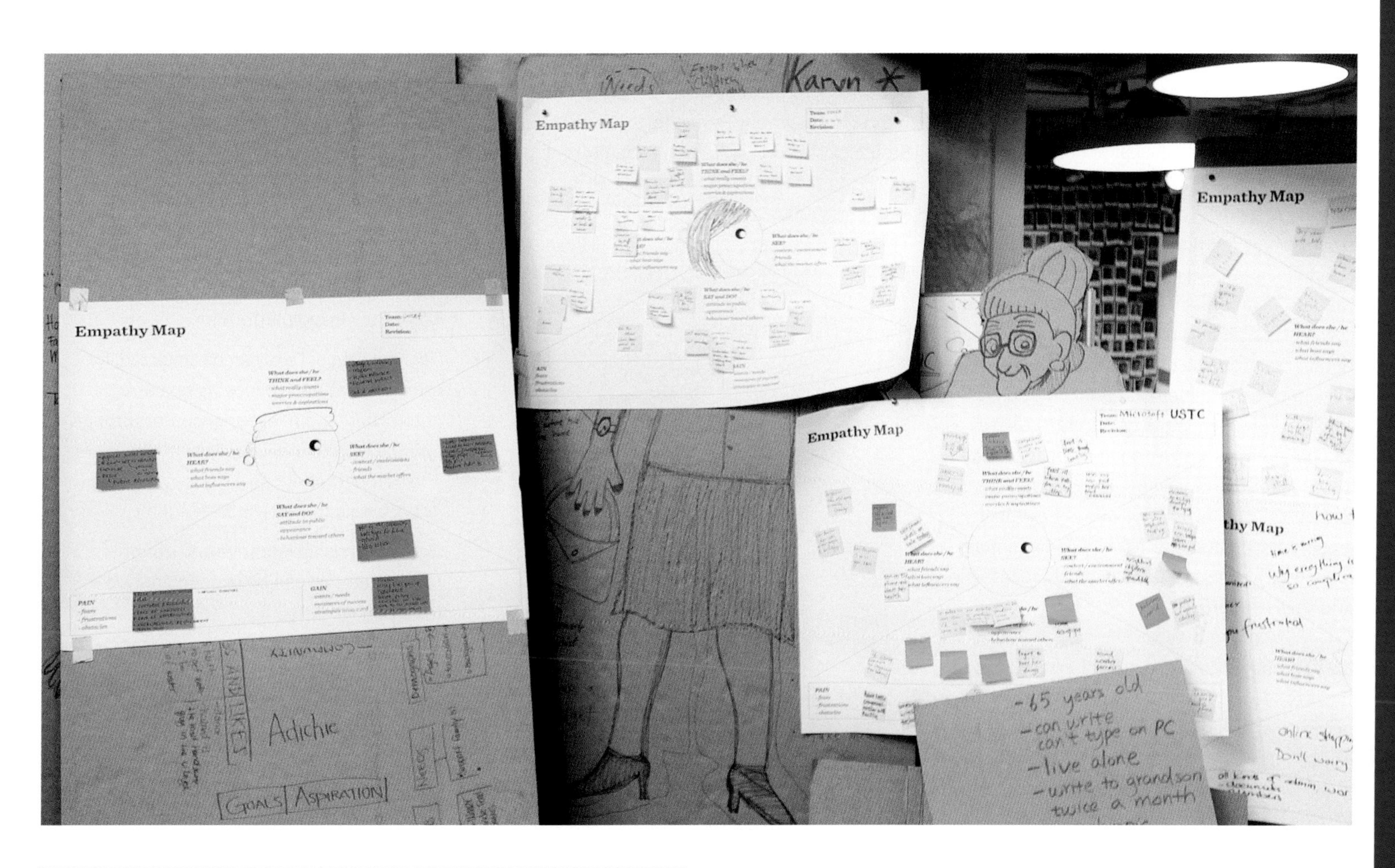

Personas and empathy maps — workshop for the SUGAR Kick-off Meeting at Stanford University (2014)

NEED FINDING AND SYNTHESIS

NETNOGRAPHY

Netnography is a combination of the word "net", short for internet, and "ethnography". Netnography describes a technique used to understand human behavior online and in the virtual world (see Beckmann and Langer 2005, pg. 221). Accordingly, data collected through ethnographic methods explores internet-based communication. Examples of material of interest include documents from online forums, online newsgroups, chats, social networks and emails. Netnography can be used for exploring associations and communities that exist only online (e.g. *Second Life* or *World of Warcraft*), as well as for communities with an online presence that also exist offline (such as discussion forums for hobbies and leisure activities or political groups). Information from these communities is generally freely available or easily accessible after registration.

Netnographers direct particular attention to the attitudes, opinions, feelings and perceptions of the given community, as well as symbols and rituals that quantitative methods could only grasp with great difficulty (see Beckmann & Langer, 2005, p. 222; Broderick & Lee, 2007).

Sources

Beckmann, S.C. & Langer, R., "Sensitive Research Topics: Netnography Revisited", 2005

Broderick, A.J. & Lee, N., "The Past, Present and Future of Observational Research in Marketing", 2007

TOOLS

- Accounts with web portals and forums such as Twitter, Facebook, and Google+

TIPS AND TRICKS

Recommended approaches for generating netnographies:

Access: Identify a community carrying out activities relevant to the question statement you are exploring, so that research leads to relevant information. Try to gain access to closed platforms.

Data collection: Data can be collected, for example, by directly copying text and notes about group experiences, or through interviews with individual group members. The netnographer can act as an observer, or become active in the community.

Data Analysis: Qualitative and quantitative methods can be used to analyze material.

Feedback to the participants: Feedback to the community explored and the members interviewed serves two purposes. On one hand, members have the chance to reflect anew on things they have said. On the other hand, the netnographer fulfills their ethical duty to participants.

NEED FINDING AND SYNTHESIS

PERSONA

Personas are idealized, typified representations of the daily lives of people within a given group. Personas serve to inspire and guide the decisions of the design thinking team. In the 80s and 90s, personas were predominantly used by marketers. Later, they were also used in developing technology and software.

Simply put, personas are descriptive models of users, customers and other interest groups. They are about building an archetype by compiling a large number of essential, common characteristics shared by a set of individuals. Personas are built from observations, interviews and focus groups. As an archetype, a persona does not represent single individuals, even though the persona's symbol is always just one person. The personification created by an archetype helps the design thinking team to directly and empathetically identify with a given group of people.

When developing new products, services and business models, personas' specific characteristics can help to differentiate segment personalities from one another. This provides a basis on which the design thinking team can later make specific development decisions, and trace these back to specific groups of people.

In tandem with the project status, the design thinking team generally delineates two types of personas:

Real size persona: Typically, the real size persona is quickly and easily constructed early in the project. The goal is to express first assumptions about the grouping of interview partners. Simple work materials are used at this stage, and the personas can be superficial.

Scientific persona: Scientific personas are well-elaborated personas, in line with the work of Cooper (2003) and others. This kind of persona emerges over a long period of time, and serves the design thinking team as a navigation point for decision-making.

Sources

Cooper, A., & Reimann, R., "About Face 2.0: The Essentials of Interaction Design", 2003, (2nd ed.)
Brenda, L. (ed.)., "Design Research: Methods and Perspectives", 2003
Silver, C., "Need Finding 101" (*PowerPoint*), 2003

TOOLS

- Lots of pictures and paper, cardboard and markers for low-resolution personas
- Word-processing program for high-resolution personas

"Persona = person + story / goal = empathy + alignment"
(Silver, 2011)

Following Cooper (2003), the approach to generating a persona can be carried out in seven steps:

1. *Base assumptions:* This first stage covers the selection of assumed typical representatives of a certain persona as interview partners, in order to review the assumptions and perceived clusters formed through observations, interviews and other methods. In many cases, at the beginning of a project there are notable discrepancies between assumptions and reality. If massive discrepancies exist, the definition of the target group needs to be adapted. The goal is to compile and stabilize the criteria used to group the interview partners.

2. *Mapping:* This stage is about understanding the interview partners' perspectives and positioning (mapping) them in relation to criteria identified earlier. The definite "correct" positioning of the interview partners within the chosen categories is less important than the "gut feeling" the design team has gained during the need finding process.

3. *Pattern:* From mapping, a pattern for an interview partners' archetype should emerge. For example, Cooper shows that people tend to be either service- or price-oriented. In this example, both categories would be good points of departure for different archetypes.

4. *Synthesis:* During synthesis, personas are fleshed out using additional data and characteristics. For example, the representation of a typical workday could include personal attitudes towards digital media, tolerance of frustrations and so on. More data should be located for each category from the previous step.

5. *Completeness and differentiation:* In this step, the design thinking team again reviews and considers the persona in regards to their completeness and differentiation. Of course, this does not mean that new knowledge gains in later project phases will not be incorporated as adaptations. Should new discrepancies arise, these will need to be corrected.

6. *Narrative:* Especially in advanced project stages, it is worthwhile creating short descriptions of personas (roughly one to two pages long). These descriptions contain important information regarding the typical day, behavioral tendencies and attitudes of the group the persona typifies.

7. *Persona type:* After all the personas have been created, the final step focuses upon prioritizing them in terms of the project at hand. Prioritization helps the design thinking team make goal-oriented decisions for the relevant groups over the course of the project. The personas of highest priority are:

a. *Primary persona:* One or more groups for whom the product, business model or service should be developed.

b. *Customer persona:* As opposed to the user of a product, focused customer personas, especially in a business environment, illustrate the requirements of this group.

c. *Negative persona:* It can also be sensible to use a persona to identify groups that would not like to use a product or service.

Wenn ich frei habe
gehe ich gern
am Rhein spazieren.
Simona, 20 Jahre
Düsseldorf

NEED FINDING AND SYNTHESIS

WHY-HOW LADDERING

"Why-how laddering" is a technique by which to glean more information from interviews. While "why" questions tend to lead to abstract answers, "how" questions help to elicit more specific answers from interview partners. Both kinds of questions are relevant. Abstract or general answers are useful for gaining an overview of the broader context and better understanding it. Specific answers deliver concrete examples of implementations or situations. The right combination of why and how questions helps to access deeper content during an interview.

TIPS AND TRICKS

This approach is easy to use, but it has to be practiced! To practice, write down the needs of a user and try to deepen your contextual understanding using "why" questions. For example: "Why does the customer have this need?" Do the same with the how question: "How do we see the customer's need?"

NOTES

NEED FINDING AND SYNTHESIS

5 WHYS

Who has not experienced children's why questions? Children ask why so often to better understand the world around them, and to make sense of events. This technique — especially in combination with why-how laddering — is used in design thinking teams. In principle, it is about the systematic investigation of problems.

In practical usage, the why question is posed up to five times. The answers help teams locate the real origin of a problem.

TIPS AND TRICKS

- Write the knowledge that needs to be explored on a whiteboard.
- Together with your team, challenge the knowledge five times with why questions.
- Write each why question on the whiteboard, along with the answers. Both questions and answers should be formulated as complete sentences.
- Use selected answer sentences from the previous question to formulate the next why question.

This method sometimes leads to trivial statements. If this happens, you either need to reconsider and rephrase the question, or you have reached the end of the exercise.

TOOLS

- Flipchart or whiteboard
- Markers

NEED FINDING AND SYNTHESIS

POINT OF VIEW

With the "point of view" or "do the pig" technique, various standpoints on the problem from a range of stakeholder perspectives are taken into account to gain new knowledge.

The various perceptions of a single pig exemplify this, and from here comes the alternative name of "do the pig". Stakeholders interested in the pig might be a farmer, a butcher, a meat-lover, a vegetarian, a veterinarian, and so on. After identifying the stakeholders, we would list the interests and perspectives of each one. While the vegetarian might perceive an exploited animal that should not be killed, the farmer might see the pig as livestock. The farmer's interests might be how to maintain the best breeding conditions. In contrast, the butcher sees the animal as meat that could result in various different products.

A concrete example of this method in a business context is the documentation of various perspectives to prepare for the introduction of a new CRM (customer relationship management) system. The CEO of a company sees the CRM system as an important measure for improving customer relations, the IT department sees it in terms of a threat to an already stretched IT budget that must now be invested in additional projects. The marketing department generally supports the introduction, as it enables the documentation of all customer contact, but the sales staff are skeptical because it makes them replaceable.

Sources

Morgan, G., "Images of Organization", 1997 (2nd ed.)

Morgan, G., "Imaginization: The Art of Creative Management", 1993

TOOLS

- Cardboard or whiteboard
- *Post-its*
- Markers

TIPS AND TRICKS

- Write the central question statement or theme for examination in the middle of a piece of cardboard or a whiteboard.
- Use brainstorming to get the team to think of as many stakeholders as possible in relation to the theme.
- Write each stakeholder on a *Post-it* and stick them all to the board around the problem.
- As soon as you believe you have identified all the stakeholders, use lines to signify the most important connections between stakeholders and the problem or theme.
- Label the connecting lines with strong words which effectively communicate the relationship between the stakeholders. The central purpose is to document the positive and negative perceptions of the stakeholders.

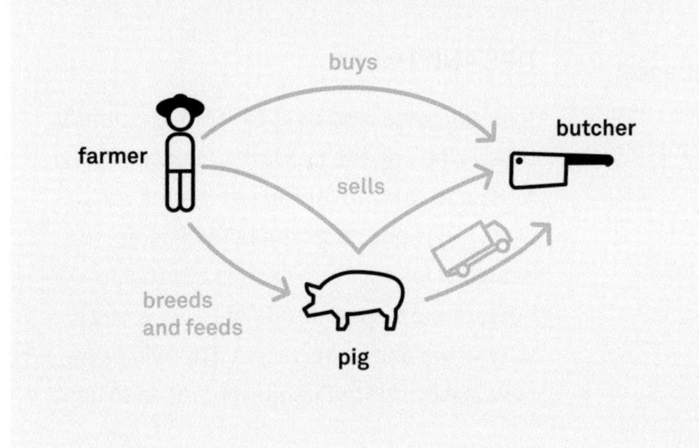

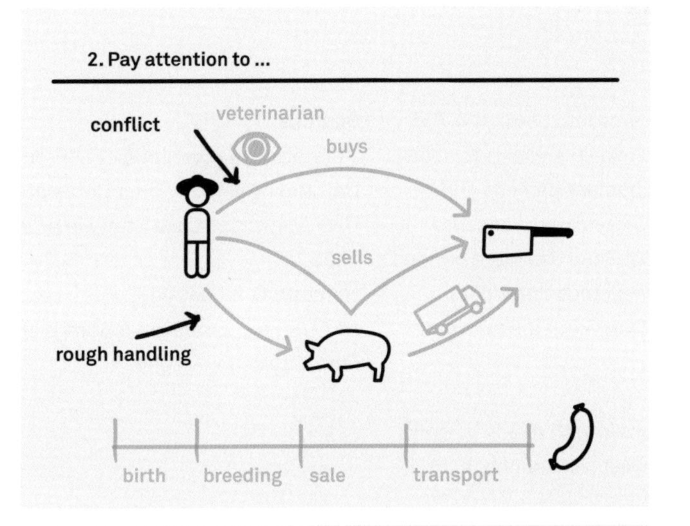

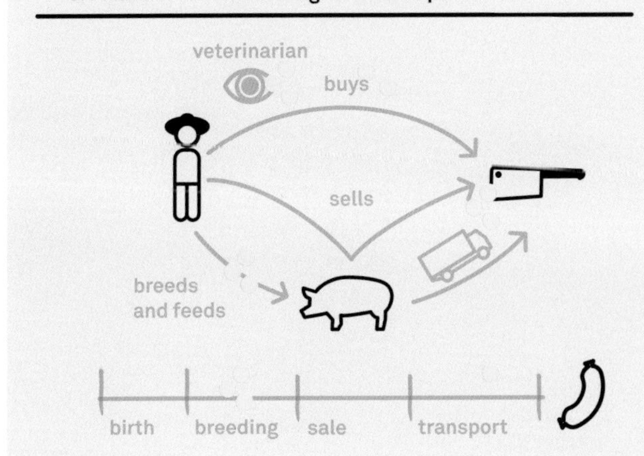

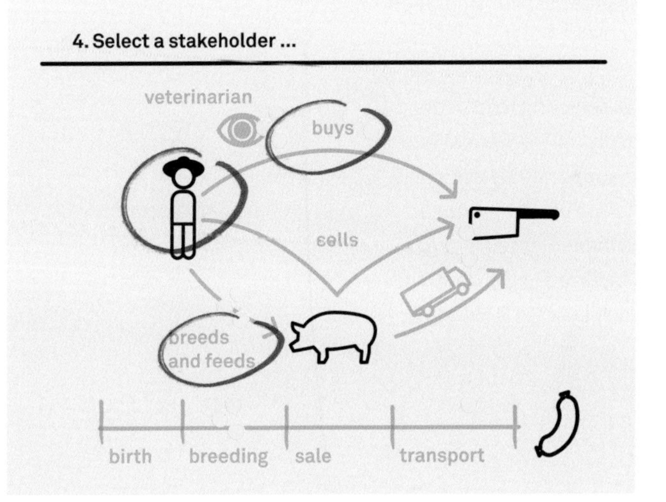

Example of “do the pig”, in reference to the stakeholders interested in the breeding and butchering of a pig

NEED FINDING AND SYNTHESIS

LEAD USER

The lead user method came into use in the 70s. With help from the lead user, the goal is to accelerate the product development process and increase the product's probability of success (see Hippel, 1986, p. 796). Lead users are defined as early users of new or extended products, processes or services. They fulfill the following two criteria:

- Lead users are confronted with needs that the majority of a large market will not notice for months, or even years.
- Lead users would have great use for a solution to the problem.

The criteria indicated for selecting lead users are confirmed by the results of theoretical and empirical research, and supported by work on the diffusion and adaptation of new products. The lead user method is implemented in organizations to reduce risks incurred in the development of new products (see Luethje & Herstatt, 2004,
p. 533).

Sources

Herstatt, C. & von Hippel, E., "From Experience: Developing New Product Concepts Via the Lead User Method: A Case Study in a 'Low-tech' Field", 1992

Herstatt, C. & Luethje, C., "The Lead User Method: An Outline of Empirical Findings and Issues for Future Research", 2004

TIPS AND TRICKS

To employ the lead user technique, identify those who use the product or service the most and take advantage of the full range of its functions. With these lead users, the design thinking team carries out the previously described methods of focus groups, interviews and observation. This will bring new, goal-oriented needs and ideas to light.

NEED FINDING AND SYNTHESIS

CAMERA STUDY

With camera studies, participants are given still cameras (digital or analog) or video cameras for a predetermined period of time. They are asked to record a specific piece of their lives with photographs and/or videos, as well as take notes and compose short descriptions of the material recorded. Smartphones offer a good alternative. Participants can immediately send pictures, videos and the commentary associated with their first impressions to the design thinking team using email or any other number of channels.

Camera studies are appropriate for a number of cases, but particularly when the presence of a researcher would lead to a change of behavior, or when his or her attendance is simply not possible. Despite the apparent efficiency of this approach, its complexity should not be underestimated, as the preparation and follow-up effort required is immense.

A concrete example of a camera study would be asking participants to take a picture of every meal they eat and, when applicable, describe the location and reason for the meal.

With the help of this technique, the team can gain realistic impressions of the participants' eating behavior while discovering meals that might have remained undisclosed during interviews.

A new medium for carrying out and supporting camera studies is Amazon Mechanical Turk. This portal, offered by Amazon, enables paid work orders (for example, for photography work) to be put online. Registered users can carry out these work orders, and are paid in accordance with previously defined fees upon successful completion of the job. Since Amazon Mechanical Turk works internationally, photos and video clips can be collected in a very short period of time. A disadvantage of this approach is that the quality of work submitted is difficult to monitor and control in comparison with a more traditional setup. Additionally, the people participating are often not available for follow-up communication.

Sources

Brenda, L. (ed.), "Design Research: Methods and Perspectives", 2003

TIPS AND TRICKS

- Review the camera or cameras to be used in the study. Ideally, a small emergency instruction kit will be provided with the cameras.
- Show the participants the functions of the cameras and hand them out.
- We recommend also distributing a small information sheet (5.8 x 8.3 inch or A5), which addresses the goal of the research and the specific situations of interest.
- While the study is under way, you can send reminders and status checks to the participants via Twitter or email.
- Collect the pictures and cameras.
- At a later date, print out the photos and discuss them with the participants who took them.

TOOLS

- Still camera, with a video function if applicable
- Twitter or email, for quick exchanges of updates and information
- Amazon Mechanical Tur

Camera study for researching the day-to-day sequence of events in households (2014). In this case, a participant from Liechtenstein took a series of photos documenting the most important events and locations over the course of a typical day.

Margarine

KAMPF GEGEN IS TERROR
US-Luftangriffe auf Ziele in Syrien

IDEATION

The ideation phase is the phase in which the design thinking team, building upon need finding and synthesis, develops ideas for potential solutions. Specialized techniques help the team to generate ideas from as many perspectives as possible.

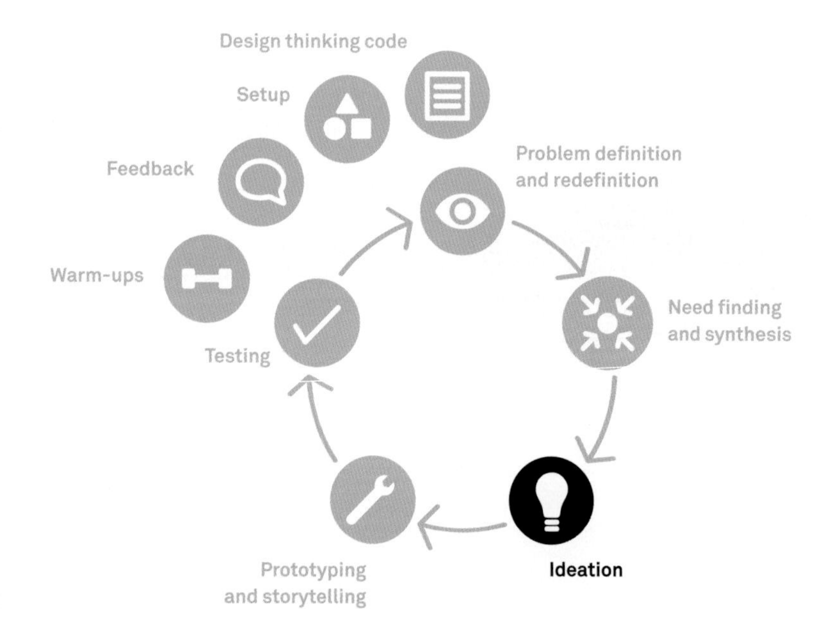

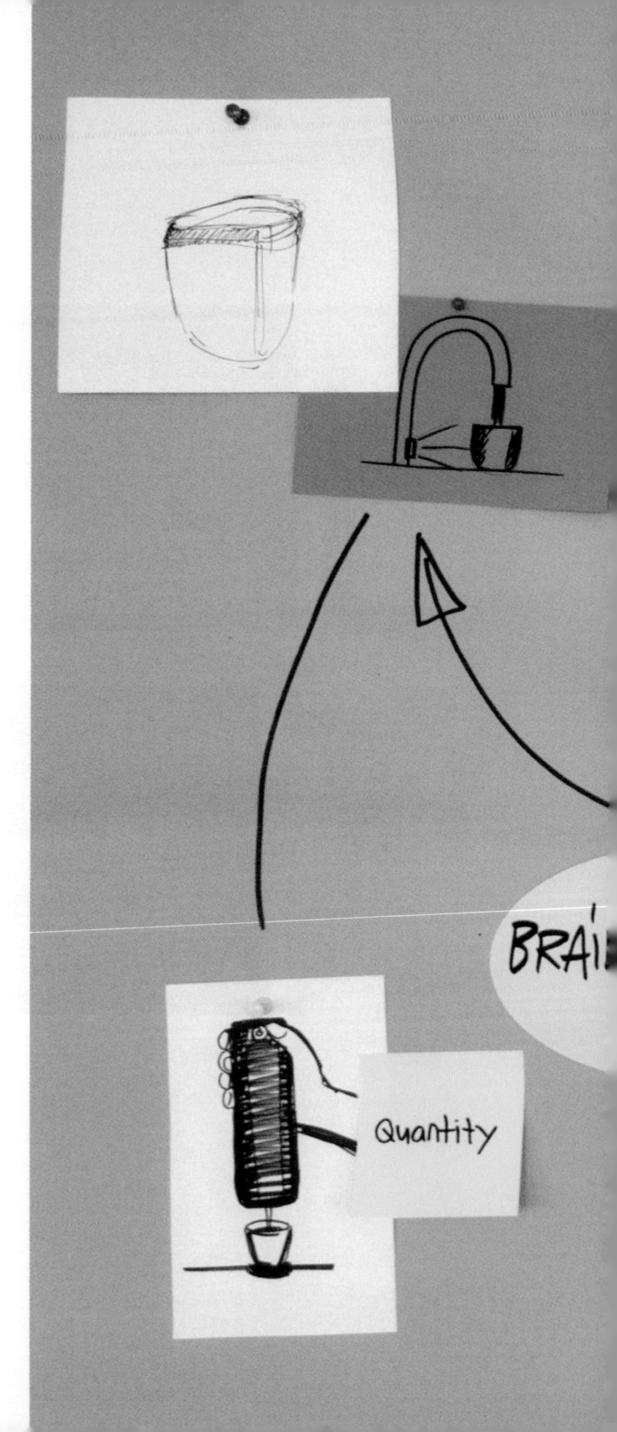

WILD IDEAS

STAY FOCUSED ON TOPIC
RMING
BE VISUAL !
target Group

IDEATION
BRAINSTORMING

Brainstorming is certainly the most best-known ideation technique, and it is a good one to consider for generating new ideas within the team. Paper and markers are enough to start the process. As a collective process of listening, knowledge sharing and idea finding, brainstorming promotes collaboration between team members. In design thinking, brainstorming is a firmly planned, methodical activity that gives the team the time and spaces to collectively begin the ideation process. Brainstorming can also be carried out spontaneously whenever the need for new ideas arises.

Brainstorming always starts with a good question statement. A question type known as a "How might we" (HMW) question helps to define the right starting point. Furthermore, there are a few ground rules to pay attention to for a good brainstorming session:

Go for quantity: The quantity of ideas is critical! The team must aim to generate as many ideas as possible in a predetermined time frame.

Defer judgment: Evaluating and discussing specific ideas should be postponed until after brainstorming.

Build on the ideas of others: With techniques such as "yes, and...", the team should try to build on ideas from their teammates before discussion moves onto the next idea.

Encourage wild ideas: Wild, unconventional ideas, which may appear ridiculous at first glance, are explicitly allowed. They help the team to take on new perspectives.

One conversation at a time: Discussions should always be focused upon one theme only. Likewise, the team should not subdivide into smaller discussion groups.

Be visual: Text is good, drawings are better. Ideas should be visualized. Even the worst drawing will be better understood within the team than text on a *Post-it*.

Stay focused on the topic: The team should ensure that they stay focused on the question statement they are trying to address.

A major challenge in brainstorming is hierarchy. Hierarchies within teams should not influence the flow of ideas and conscious divergence of answers.

TIPS AND TRICKS

- Chocolate and candy!
- Every participant should have *Post-its* and a marker to hand.
- Write the brainstorming question on a large piece of brown paper, cardboard or whiteboard.
- Every participant may contribute his or her ideas. Try to stick to the brainstorming rules!
- As soon as a team member has contributed an idea, the team should try to elaborate it using the "yes, and..." technique.
- Once an idea's potential has been fully elaborated, another team member may contribute a new idea.

The brainstorming technique sounds easy. However, coming up with many good ideas in a short period of time requires a lot of practice within the team. If the team is tired, they can do a warm-up beforehand.

1

2

3

TOOLS

- Markers (ideally Sharpies or Edding 1200)
- Stopwatch
- Gong

4

5

Design thinking bootcamp with students from the University of St. Gallen Master in Business Innovation (MBI) and MBA programs at the Swisscom BrainGym in Bern. The BrainGym offers Swisscom employees the ideal space in which to carry out human-centered design methods.

IDEATION
BRAINWRITING

Brainwriting is a further development of brainstorming in which each participant writes three ideas for a question statement on a piece of paper. These idea notes are passed to neighboring participants and further developed or completed in the course of five rounds.

To prepare for brainwriting, the question statement is discussed. In this process, only questions of comprehension are allowed; solutions are not discussed. This preparation helps to sharpen the group's understanding of the goal. At the end of a brainwriting session, there is a rich well of ideas for the group to discuss and prioritize.

The advantage of brainwriting is that hierarchical issues that might influence individual group members' thinking are shut out. Hauschildt argues that brainwriting leads to a similarly high performance to brainstorming.

TOOLS

- Foamboards or cardboard on which to put up results
- Markers (ideally Sharpies or Edding 1200s)
- Stopwatch
- Gong

Sources

Hauschildt, J., "Innovation management", 2004

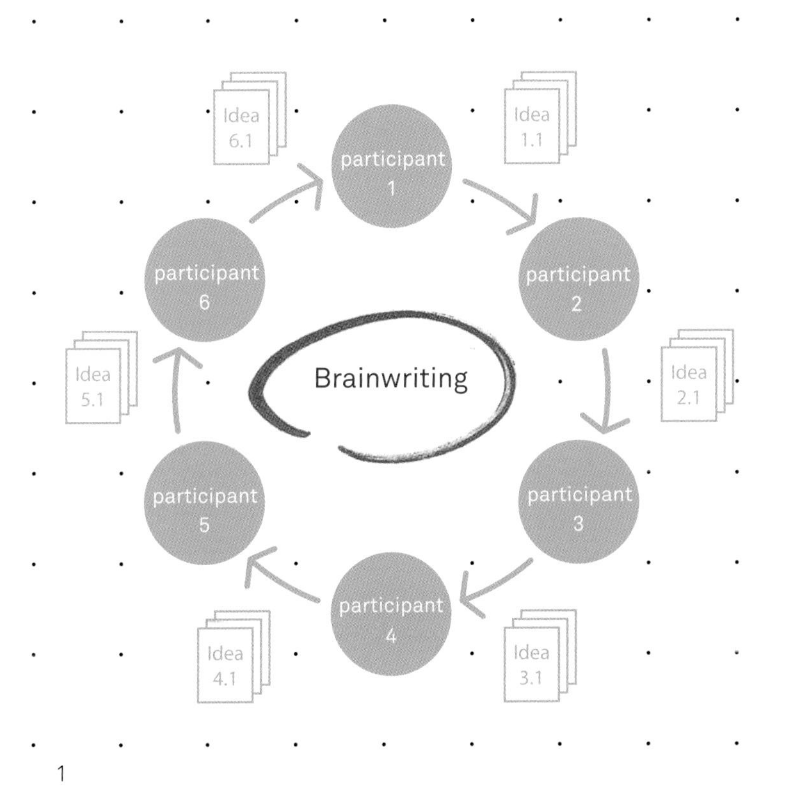

1

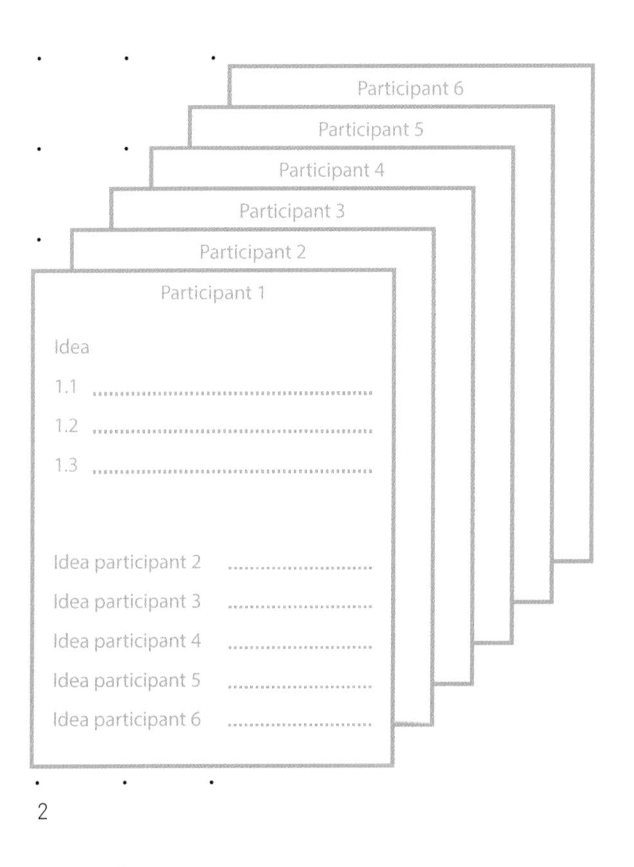

2

1 Illustration of the brainwriting technique (Hauschildt, 2004)

2 Example of a brainwriting template

IDEATION

SIX THINKING HATS/LATERAL THINKING

The "six thinking hats" or "lateral thinking" are techniques by which problems and questions can be considered from different perspectives in a systematic, goal-oriented manner. Edward de Bono popularized this concept.

The six thinking hats characterize the six perspectives that should be considered when examining a problem, according to de Bono. In its application, each team member gets a hat and with it, a specific frame of mind and perspective which with to enter the discussion. De Bono defines the following thinking hats:

White hat: The wearer takes a fact- and number-oriented perspective regarding the question, embodying a neutral, matter-of-fact and objective temperament. Only facts count — gaps in the data are quickly recognized and named.

Red hat: The wearer has a very emotionally-driven perspective, which goes so far as to be characterized by rage. The goal is to experience and evaluate emotional reactions.

Black hat: Negative and risk averse, this point of view tries to identify and address problems and weaknesses.

Yellow hat: The wearer of this hat is positive and optimistic. The goal is to identify all benefits and advantages of a potential solution.

Green hat: The green hat represents new ideas and creativity.

Blue hat: Cold and focused, this point of view focuses on the management, control and organization of the group's ideation process.

Sources

De Bono, E., "Six Thinking Hats", 1999 (2nd ed.)

TIPS AND TRICKS

- Allocate a hat to each team member and once more describe the role of each hat.
- Discuss the given issue within the team. Each participant takes on the role assigned by his or her hat during the discussion.
- Each participant should take their own notes on the discussion's results.
- Following this exercise, participants should discuss and reflect upon the results. The participants should use their personal notes for this part of the process.

MATERIALS

- Notebooks
- Colorful, possibly even funny hats (optional)
- Video camera to record the interaction, and enable later analysis (optional)

De Bono's six thinking hats (1999)

IDEATION

POWER OF TEN

The "power of ten" technique can be used in both the ideation and synthesis phases of need finding. It enables the team to contemplate a problem on different levels — both larger and smaller than may be the case in reality. If a team finds itself in a brainstorming process that is not yielding useful results, the power of ten can establish restrictions that increase creativity. For example, "How might a solution look if the team had a million dollar budget?" or, "How might the solution look if the team had a 50 dollar budget?"

Of course there are no limits to what expansion and reduction criteria can be applied.

Sources

d.school Stanford University, 2015b

TIPS AND TRICKS

- Write the initial question statement on a piece of cardboard or a whiteboard.
- As a group, list all the possible properties and restrictions of the question statement.
- Within your team, select the interesting properties and restrictions, and expand or reduce them.
- Reformulate the initial question statement according to the newly expanded or reduced dimensions selected.

MATERIAL

- Cardboard
- Markers
- *Post-its*

IDEATION
HOW MIGHT WE

"How might we" question statement formulation (HMW) is one of the most important techniques in design thinking, and in the ideation phase in particular. The correct method of asking questions tends to enormously increase the quality of results in subsequent techniques. The question "how might we" helps prevent assumed solution parameters influencing the question statement. This allows and inspires more effective consideration of potential solutions.

Here are two examples:

1. How might we deliver healthy food to our customers as quickly as possible?
2. How might we improve the training conditions available to sporty members of the workforce?

Sources

d.school Stanford University, 2015b

TIPS AND TRICKS

Use HMW questions to guide brainstorming whenever possible.

- Write "how might we" on a whiteboard, flipchart or packing paper.
- Discuss how the HMW question statement might be formulated within the team. Write out alternative suggestions, and evaluate them together.

NOTES

PROTOTYPING AND STORYTELLING

Prototyping is a technique that simulates real products, services and business models at the end of design thinking projects. The goal is to make a prototype that an R&D department can implement as soon as the design thinking project finishes. Naturally the combinations and creativity of the prototype variations listed here are unlimited, so that other prototype forms can also take shape.

TIPS AND TRICKS

Be fast: Try to build quickly in the early stages of a design thinking project! Prototypes should not require lots of time and money.

Stay open: Even when you think you have found the final answer with your ideas and prototypes, stay open to changes as well as completely new ideas.

Good enough: Adapt the prototype generated in each project phase so that it is just good enough.

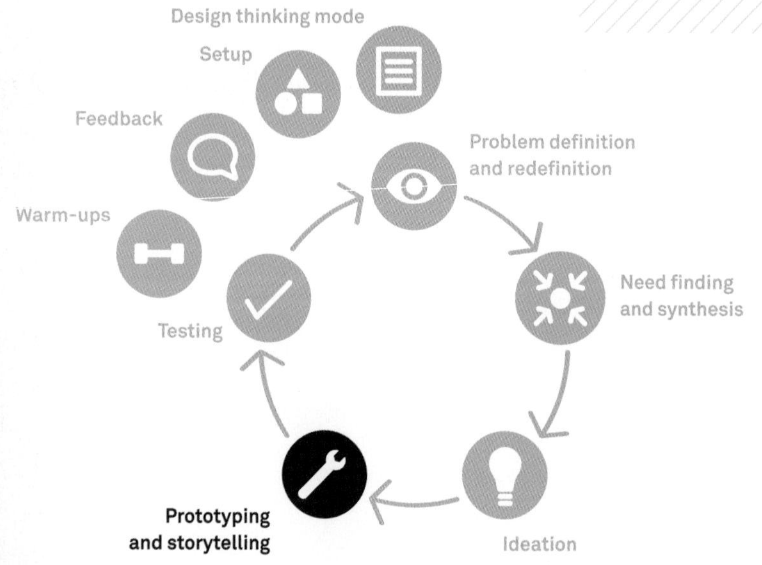

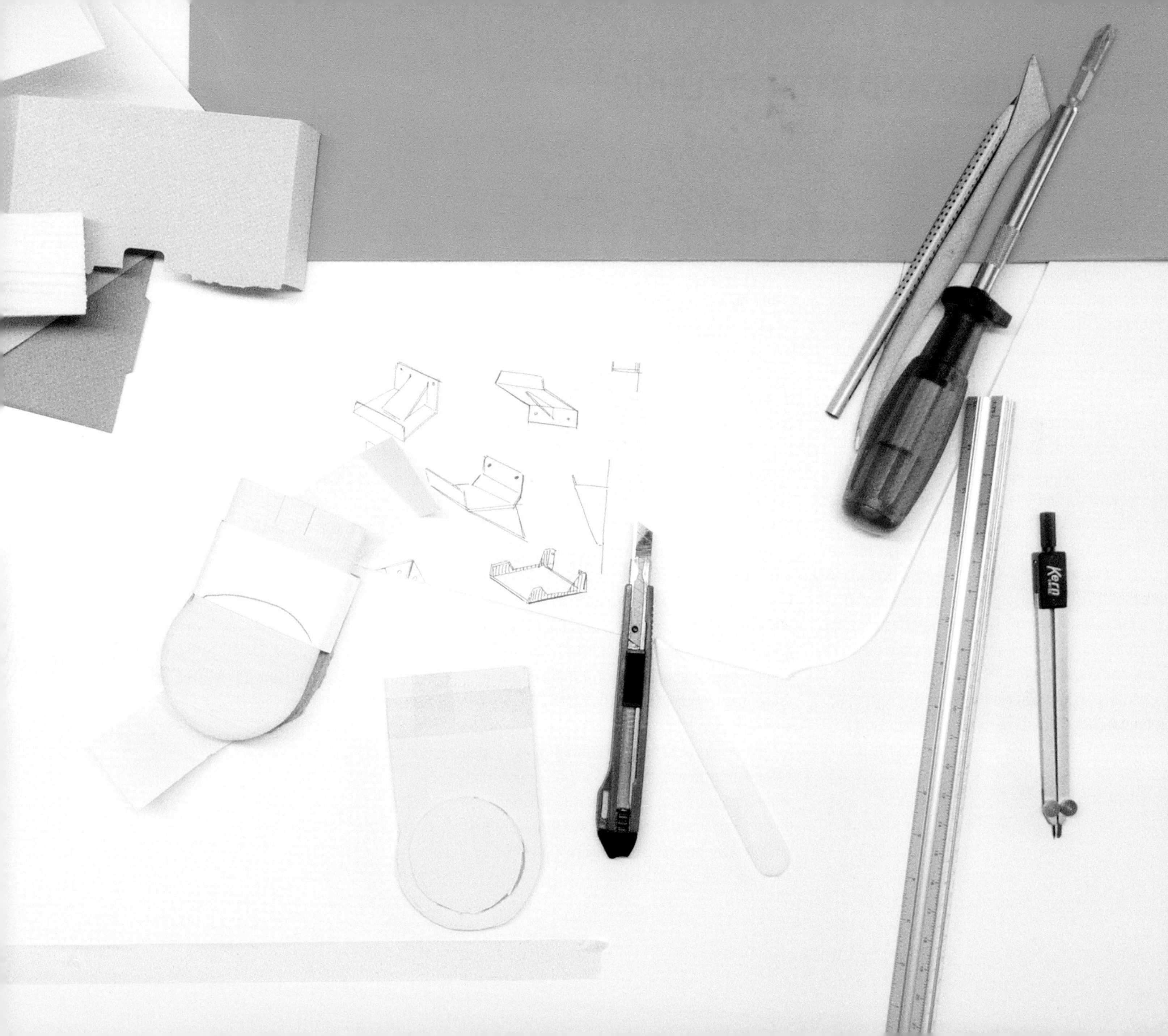

PROTOTYPING AND STORYTELLING

PROTOTYPING

Prototyping does not first begin after weeks of research; by way of the micro-cycle it begins as soon as a design thinking project starts. But how are these first prototypes built? The principle is that early prototypes should be low-resolution prototypes (LRPs), while high-resolution prototypes (HRPs) follow in later project phases.

Low-resolution prototypes: LRPs enable underlying principles and single functions to be efficiently verified via tangible expression and communication outside the team during the early project phases. In engineering, focused prototypes are spoken of. LRPs are generally rudimentary in appearance. They are however quick and inexpensive to build; some prototypes can be built in just a few minutes. Through this prototyping method, the rudiments of ideas can be quickly tested with clients and users, and failures or misinterpretations identified and discarded. For example, we built numerous low-resolution prototypes focused on increasing the degree of automation in laboratory equipment for a pharmaceutical company. These prototypes not only helped the team to quickly achieve an understanding and empathy for the complex and challenging everyday happenings in laboratories, but also communicated their own ideas. Another advantage of LRPs is that the danger of falling in love with your own work and no longer searching for new ideas is much lower than with HRPs.

High-resolution prototypes: In every design thinking project, HRPs are also built. HRPs are well thought through from technical and economic perspectives. In some projects, these prototypes are already so close to reality that users think that the product or service is already available to buy. For example, in 2011 we built a fully integrated solution for a car-sharing system for an automotive company. The prototype could already open, shut, start and stop the car via a smartphone. Using a web portal connection, reservations could be made and pricing models established. In essence, HRPs are substantially more demanding to build and require more investment than LRPs. HRPs are therefore first built in later phases of the project (beginning with the funky prototype).

In this model, the number of prototypes correlates with both the degree of prototype resolution and project phase. Many prototypes are built in the critical function, dark horse and funky prototype phases. True to the motto "fail early and often", a high number of early prototypes serves to test the team's assumptions and ideas with users and other stakeholders. Using this approach, correct assumptions are quickly validated and those that are incorrect are quickly identified and eliminated. Beginning in the funky prototype phase, the number of prototypes is explicitly reduced. At the same time, the resolution of the prototypes is increased, as is the effort invested in them.

Sources

Ulrich, K.T. & Eppinger, S.D., "Product Design and Development", 1995 (4th ed.)

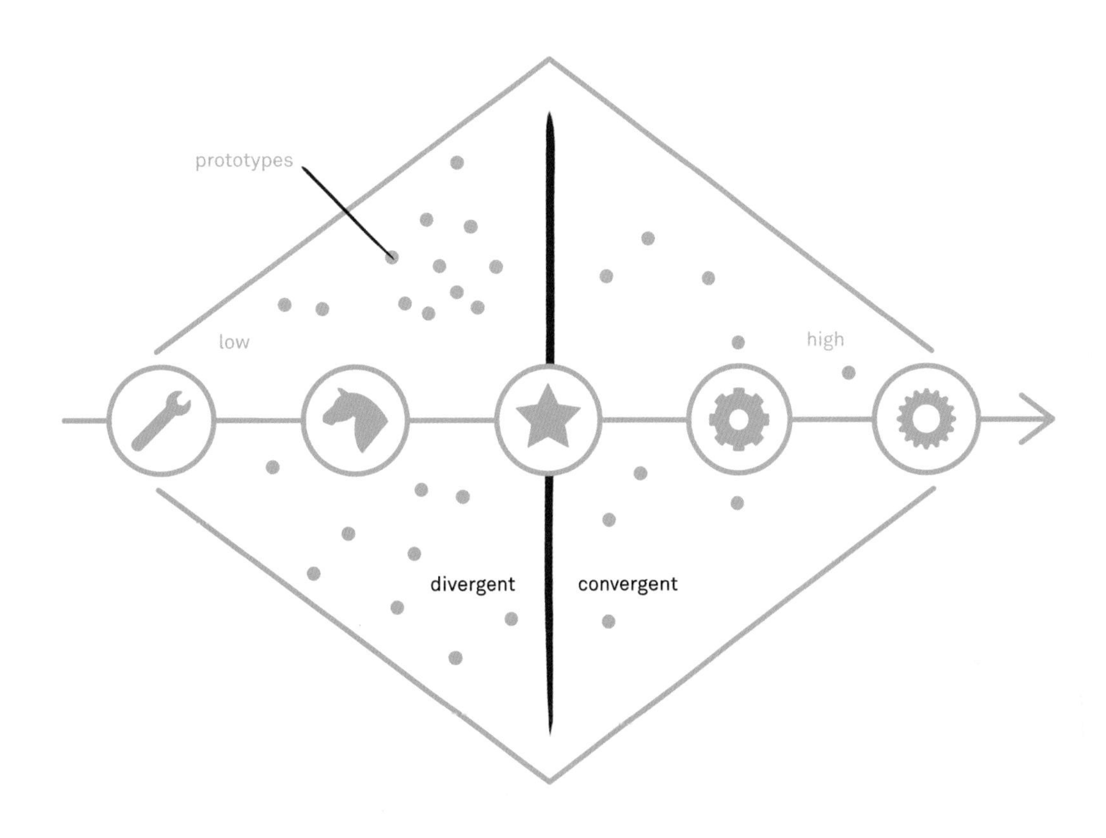

"Prototypes evolve from low resolution to high resolution."

PROTOTYPING AND STORYTELLING

WIREFRAMING

A wireframe is a rudimentary, functionally reduced demonstration of user interfaces on computers or other devices such as smartwatches, smartphones, and so forth. Wireframes are used to show buttons, notifications, general principles of navigation and/or elemental sequences. Through this, the design team can collect focused feedback about usability, order, and sequences through testing.

The goal of using wireframes is to make functions concrete as quickly as possible through prototyping, and prepare them for testing with users. Even simple drawings on *Post-its* can be put to use. However, other materials, such as paper or cardboard, are also popular in the early phases of a design thinking project. For example, we simulated an interactive windshield for an automotive company. We did this by printing various pieces of information over a landscape onto a self-adhesive sheet.

From a purely practical perspective, the advantage of wireframes is that through their simplicity, they give users a feeling that their feedback can have a real impact. In contrast, a perfectly elaborated mock-up can inhibit users giving feedback, as they may perceive the product or service to be already complete. This renders their feedback irrelevant. Simplicity makes wireframes perfect tools in the early phases of a project.

A tool producer Balsamiq has already made something of this knowledge. Balsamiq software generates user interface sketches that look hand-drawn. Similar tools for mobile devices are POP and Prott.

TIPS AND TRICKS

- Use simple materials such as paper and cardboard.
- Perfectionism does not help you here; wireframes should never be too complicated.
- Modular variations of wireframes have often proved useful, as users can, for example, configure their own agendas.

TOOLS

- https://popapp.in
- https://prottapp.com
- https://balsamiq.com

1

2

3

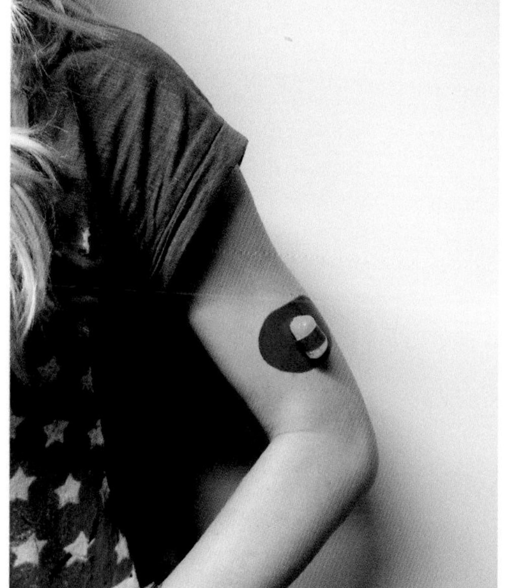

4

5

1 Making a wireframe with the help of *Post-its*
2 Simulation of data glasses
3 Wireframe on a mobile phone
4 Prototype of an emotion sensor intended to help banking consultants assess their clients' moods
5 Hand-drawn wireframe on a laptop (the drawings were fixed one after the other on a screen to simulate the process)

PROTOTYPING AND STORYTELLING

MOCK-UPS

Mock-ups are a preview of a computer system's user interface or other product or service systems. Representations are developed to be as realistic as possible. The value of mock-ups lies particularly in testing users' overall impressions of a system. Through the highly realistic appearance of the simulation, the users and customers with whom it is being tested can be asked very specific questions.

Substantially fewer mock-ups than wireframes are built, since the production of mock-ups is far more demanding. As a result, tested wireframes are turned into mock-ups during later project phases. Products such as *Photoshop* and *Illustrator* are often employed as tools. That said, design thinking teams also often use simple HTML 5.0-based mock-ups.

An example: After testing numerous wireframes for a European software producer, we developed a mock-up for a web portal that aimed to improve communication between suppliers and producers in the chemical supply chain (B2B). With the help of this prototype, the whole product could be shown realistically. However, teams should be ensure that mock-ups are not used too early, but only after wireframes have been built and tested.

TIPS AND TRICKS

Horizontal, not vertical: With mock-ups, limit yourself to surface-level development. Do not concentrate upon the back-end functionalities upon which they are based.

SOFTWARE

- Adobe Photoshop
- Adobe Illustrator
- HTML 5.0 editor

1

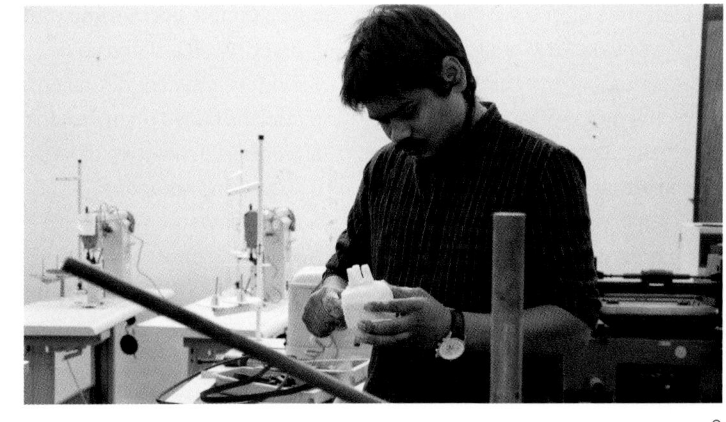

2

3

1 High-resolution mock-up to show passenger car data on a mobile phone in real time
2 Working on a physical mock-up. Collaborative project between the University of St. Gallen and the Indian Institute of Technology Kanpur (IIT; 2015)
3 Mock-up to simulate notification of a mechanical failure on a smartwatch

PROTOTYPING AND STORYTELLING

OPEN HARDWARE

Open hardware platforms offer a simple, inexpensive and quick way to build prototypes with extensive sensory, actuating and computing elements. Worldwide, the best-known platform is *Arduino*. It was initially conceptualized for the do-it-yourself community, but even large organizations are increasingly using these platforms as resources for efficiently generating middle- and high-resolution prototypes. *Arduino*, for example, offers a series of analog and digital interfaces that can be connected to a variety of sensors and motors. With only minimal knowledge of the C and C++ programming languages, relatively simple electromechanical systems can be built. Videos can be found online showing everything from automatic feeders for cats to automatic watering systems for every use imaginable within the home.

The advantage of the open hardware platform is that it is compatible with a large number of devices. This eliminates the need for the complex calculation of resistance, circuits or connection soldering. Instead, the team can work with standard components on a printed circuit board and connect them using plugs and sockets.

Some example portals and platforms are:

- Arduino: www.arduino.cc
- Sparkfun: www.sparkfun.com — offers own platforms based upon *Arduino*, as well as an extensive selection of sensors, switches, displays and actuators.
- Adafruit: www.adafruit.com — similar to Sparkfun, it offers components from its own product line. The whole portfolio is well documented and also appropriate for beginners in the field.
- LittleBits: www.littlebits.com — also based upon the *Arduino* architecture. These components can be combined and programmed even more easily and with less effort.
- Electric Imp: www.electricimp.com — extremely small, energy saving platform that can be programmed with the help of Squirrel (a programming language based on C) that has a WiFi connection as standard.
- Raspberry Pi: www.raspberrypi.org — well-known platform that contains a complete computer processor with graphics, multiple USB connections and ethernet connectivity.
- Fritzing: http://fritzing.org/home — learning platform built upon *Arduino.*

TIPS AND TRICKS

Curiosity: Just try once to build your own hardware. LittleBits offers an easy entry for everyone.

Beginners: Look around online and perhaps take a workshop for beginners as a group.

Modules instead of integration: Try not to combine all modules with one another at first. Instead, build manageable modules and simulate the integration of the individual components. This greatly simplifies the entry level.

1

2

TOOLS

- Micro-controller or printed circuit board
- Software for programming
- Accessories for programming: Programmer, USB cable, etc.
- Soldering iron
- Electronic components: resistors, condensers, transistors, LEDs, sensors, switches, etc.
- Connecting wires, breadboards, side-cutting pliers, alligator clips
- Additional components such as housings, chargers, etc.

1 Sensor and display to determine the soil moisture of house plants using the *Arduino* platform

2 Prototype to determine an athlete's VO_2 max (maximum oxygen uptake) with the help of the *Arduino* platform and CO_2 and O_2 sensors

PROTOTYPING AND STORYTELLING

ROLEPLAY

Roleplay is a technique particularly appropriate for simulating and prototyping services and business models. In roleplay, at least one actor takes on the role of a predetermined character and plays it out. The goal is that audience members, or those taking on the tester role during the roleplay, can experience the most realistic possible perception of the situation.

The advantage of roleplay is that it needs only a relatively manageable material effort to make it possible. In addition, roleplay can be used with both low- and high-resolution prototypes. In many design thinking projects using roleplay, small scripts with directions and guidelines for the characters are written in advance. For example, for a software company we used roleplay to demonstrate numerous consulting scenarios between a tax consultant and his client in order to determine which new approach to software support might work best and gain acceptance in the future.

Next to planned and scripted roleplay, there are situations in which improvised roleplay is used. We primarily use these variations to generate the scripts for planned roleplay. Through this, we generate what we might call prototypes for prototypes.

Critics of the technique warn that people are often not in the position to take on a given character. This critique can be observed in practice. However, from our perspective, this fact is not sufficient reason to forgo roleplay, as long as the central functions of the prototypes remain preserved. However, if the team falls short of this objective, perhaps it should shift to another technique, such as comics.

TIPS AND TRICKS

We recommend that the team develops the story together on a whiteboard, ensuring the key phrases are retained. Then, it is just practice, practice, practice!

Roleplays simulating workflows and situations to demonstrate the interaction of machines, software and people, University of St. Gallen (2013)

PROTOTYPING AND STORYTELLING

BODYSTORMING

Bodystorming is a specific roleplay technique in which one's own body is improvisationally used to allow situations to be felt. Bodystorming is both a prototyping and a brainstorming technique, through which ideas can be tested, but new insights can also be won.

An example is the simulation of an intercontinental flight in a ground-based imitation of an airplane cabin. Without actually using an airplane, the functions of an existing or newly developed interior can be simulated and tested. The advantages of substantially reduced costs and time requirements are clear.

As preparation for a bodystorming session, team members typically observe and analyze the situations they intend to imitate. The results are used to direct specific questions for the bodystorming. If new dimensions and improvements are to be added, these should clearly also be considered. The building of the situation then follows or, if possible, the participants report directly to the location. There, bodystorming is either used as an instrument of brainstorming, or carried out as part of testing with clients.

Success criteria for bodystorming are:

Briefing: The bodystorming and testing participants must be comprehensively informed about the course of events in advance.

New users: People that enter a new situation for the first time often require more time to orient themselves to the circumstances. Accordingly, this should be considered in project planning.

Similarity: If possible, when a new testing environment is selected, it should be done with consideration to the strongest possible similarities to real situations. If environments do not feel close to reality, they can lead to false results.

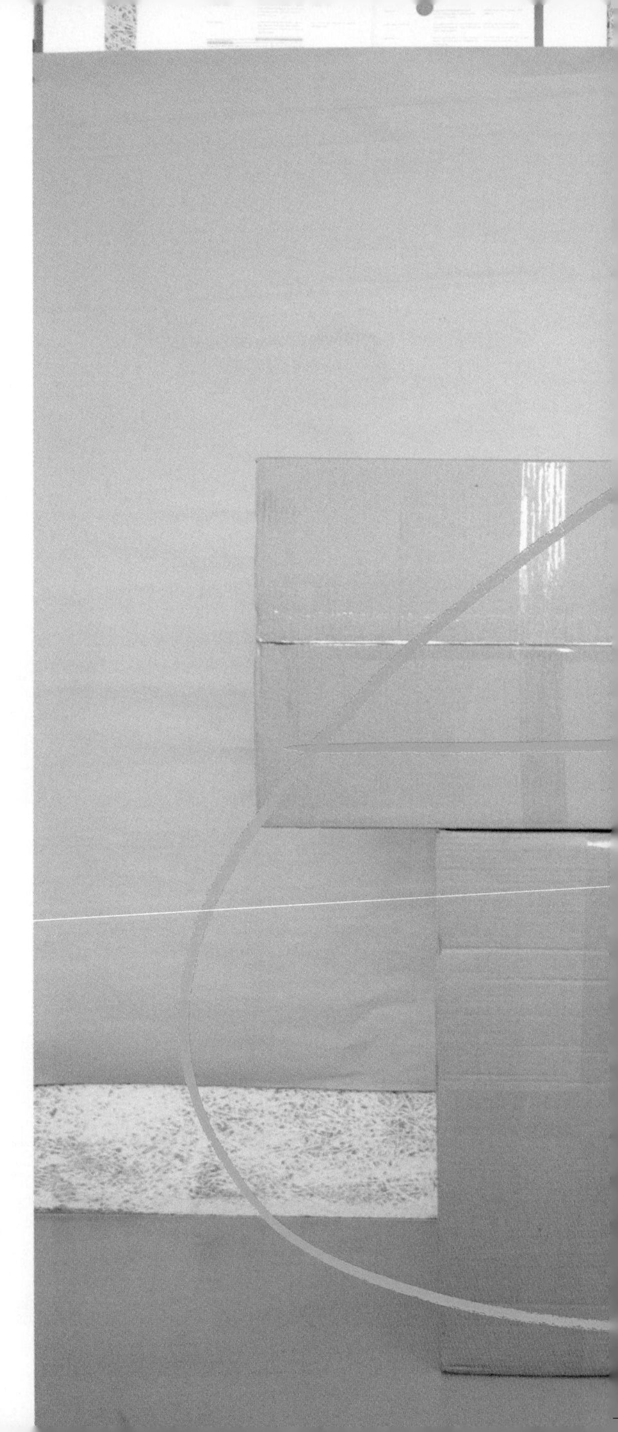

PROTOTYPING AND STORYTELLING

PAPER PROTOTYPING

Building prototypes out of paper is a powerful tool. Unfortunately this is often underestimated. Starting with user interfaces in computer programs, going all the way to entire automobile cockpits, we have built hundreds of prototypes out of paper and cardboard. At the outset of the University of St. Gallen and Stanford design thinking programs, students always have to build bicycles out of paper (other materials are restricted to a total weight of 500 grams). These paper bikes must carry at least one person, and survive a previously communicated tactical game.

The space for using prototypes made out of paper is not limited to the representation of digital material only, but extends into the analog fields of our lives! For a company in the chemical industry, we easily simulated a quality seal on packages with the help of simple paper stickers. In another case, we built a "health meter" out of paper. This was a screen for attachment to one's arm which was intended to measure and display the individual's health status.

The advantages of this approach are clear: in contrast to digital approaches, the cost and effort of developing prototypes is extremely low. Prototypes made out of paper are therefore especially appropriate in the first phases of a design thinking project, but can also be built later on.

MATERIALS

- Markers for drawing and writing
- Wire and twine
- Scissors and sharp knives (e.g. box cutters)
- Styrofoam boards

TIPS AND TRICKS

"Start doing it"— motivate your team to build prototypes out of paper, even if it appears too trivial at first.

1

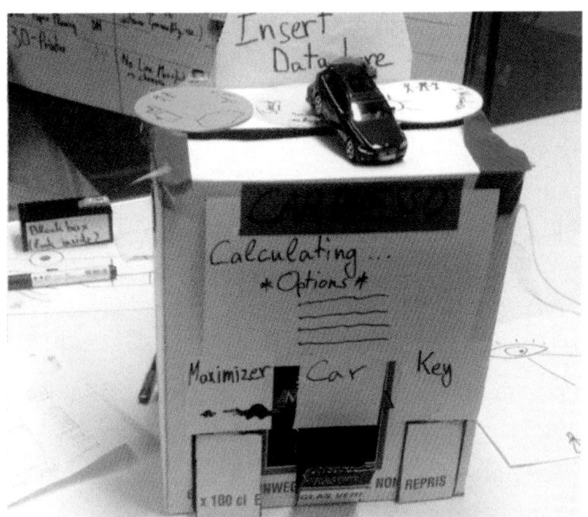

2

3

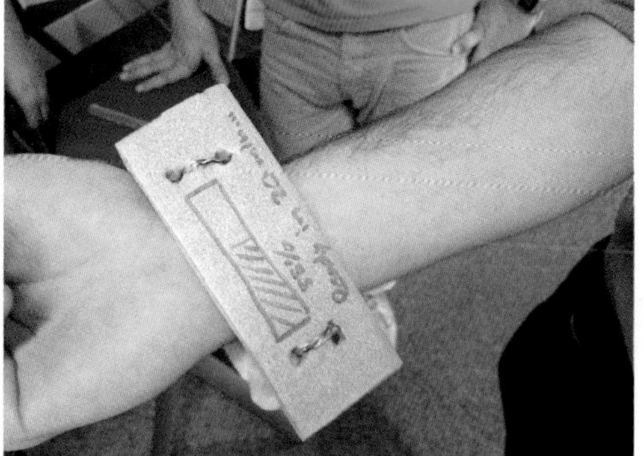
4

1 A sticker on a spraycan to simulate a quality seal
2 The “Carpresso” simulates the output of an automobile depending upon the user’s mood or desired driving behavior for the car (sporty, elegant, comfortable)
3 Simulation of a chip in a sneaker to show movement data
4 Prototype of a wrist sensor to display various medical body measurements

PROTOTYPING AND STORYTELLING

STORYTELLING AND STORYWRITING

Storytelling is a technique enabling the communication and presentation of various scenarios and stories. In contrast to roleplay, with storytelling there is a teller instead of a group of actors. The storyteller interacts with the listeners while telling the story, and is thereby able to individualize the flow of the storytelling in parts.

The disadvantage of the storytelling variant is that the quality of the prototype depends primarily upon the public speaking abilities and personal presence of the storyteller. Not all people are made for storytelling, and in actual situations, this should be considered. This approach calls for honesty between team members.

In addition to verbal storytelling, stories can also be communicated through writing, pictures, animations and videos:

Written stories: Written stories do not demand that the teller and listeners be present in the same place at the same time. This method is particularly useful when teams need to test prototypes across continents. After a description is written down once, the prototype can be sent to the tester with virtually no further explanation.

Visual stories: Visual stories are pictorial stories with or without text. The pictures directly increase the story's comprehensibility. However, visualizations cannot be generated for every prototype. Comics are another technique that will be discussed separately.

Animation and videos: Animation and videos offer many possibilities to tell stories audio-visually. Like videos on websites such as Kickstarter, the idea is clarified and a prototype shown. Producing this kind of prototype requires effort, which is the disadvantage of this method. Because of this, we generally only use this technique when building high-resolution prototypes.

TIPS AND TRICKS

- Develop a storyboard within your team, that is, a chronological run-through of the scenes to be demonstrated. A whiteboard is usually suitable for this.
- Essential phrases should be recorded in writing for the primary speaker and each role.
- Practice telling the story multiple times and use the feedback techniques described in this book.

Storytelling at a company workshop at the University of St. Gallen (2014)

PROTOTYPING AND STORYTELLING

COMICS

Comics are a subcategory of storytelling and offer the design thinking team a very concrete and often funny method of presenting a product, service or business model with the help of a small story. Our experience has been that comics have proven themselves for prototypes of services and business models. While just a few years ago it was necessary to have someone in the team who could draw comics in order to take advantage of this approach, teams can now take advantage of the numerous tools available online for drawing comics, and even generating animations.

The advantage of comics is that they reduce contact barriers, particularly with younger groups. Because comics are amusing, testers in the resulting discussions are often more engaged and open to conversation than with other methods.

TIPS AND TRICKS

- Before you begin drawing comics, think about the chronology of the story within your team. You can use tools such as a whiteboard to develop the story.
- Within the team, develop a key picture for each scene that best expresses your statement
- Draw a first rough draft of the comic and test it within the team.
- Integrate the team feedback and refine the comic.

TOOLS

- www.wittycomics.com
- http://bitstrips.com
- www.stripcreator.com

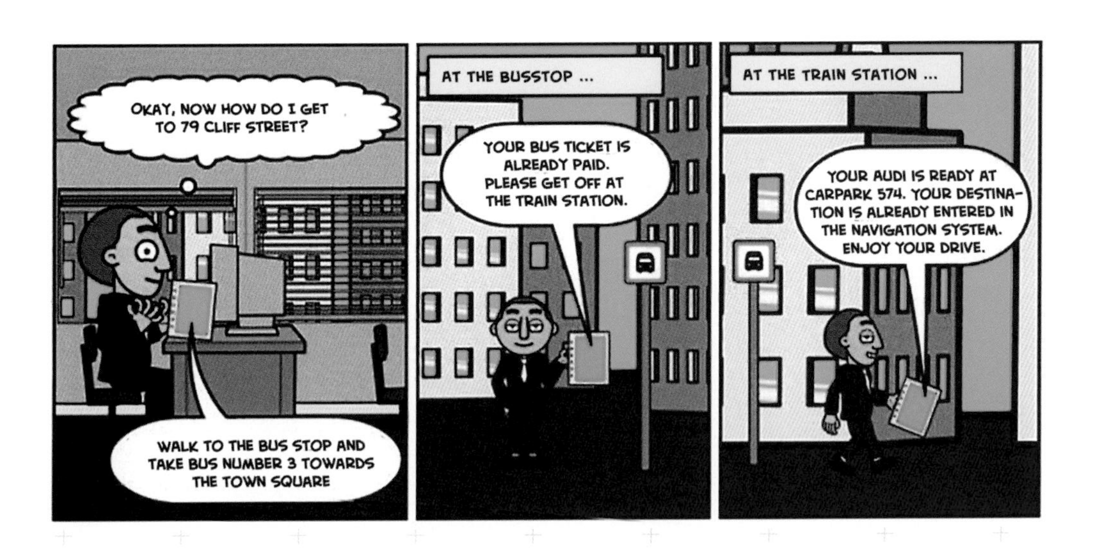

Demonstration of a new mobility concept to navigate people to a goal destination using different transportation methods (train, car, bus); prototype developed at the University of St. Gallen, 2012

PROTOTYPING AND STORYTELLING

RAPID 3D PROTOTYPING

Three-dimensional printing has generated real hype over the past few years. Today, it seems that almost all objects can be printed using this method — from simple components such as plugs and sockets, to fashion accessories and art, to tissue structures for the human body and components for houses and airplanes. What these operations all have in common is that the complexities of creating shapes and tools are reducing dramatically. Instead, materials like plastic, cement, conductive materials and tissue substitutes are three-dimensionally layered, or manipulated to harden by laser.

For design thinking teams, this is a great approach for all forms of middle- and high- resolution prototypes. Numerous projects by young founders have ensured that there are a large number of inexpensive alternatives to the professional devices which are generally only acquirable for for a five-figure sum. With online service providers of 3D printing such as shapeways.com and i.materialise.com, only the data needs be uploaded and the final printouts, offered in a variety of materials, can be received via mail. However, local associations such as FabLabs also have devices in stock, including 3D printers and laser cutters, that can be used upon request. Finally, private owners of 3D printers often take print orders. Many are registered on the 3DHubs community for this exact purpose.

Sources

www.shapeways.com
www.i.materialise.com
https://en.wikipedia.org/wiki/Fab_lab
www.3dhubs.com

TIPS AND TRICKS

The easiest way to get access to a 3D printer is to visit FabLab. Look at the technology and talk to the experts!

TOOLS

- Google SketchUp — quick and easy development of 3D models
- 3D printer, e.g. www.makerbot.com
- Additional materials such as filaments: https://www.3dware.ch/en

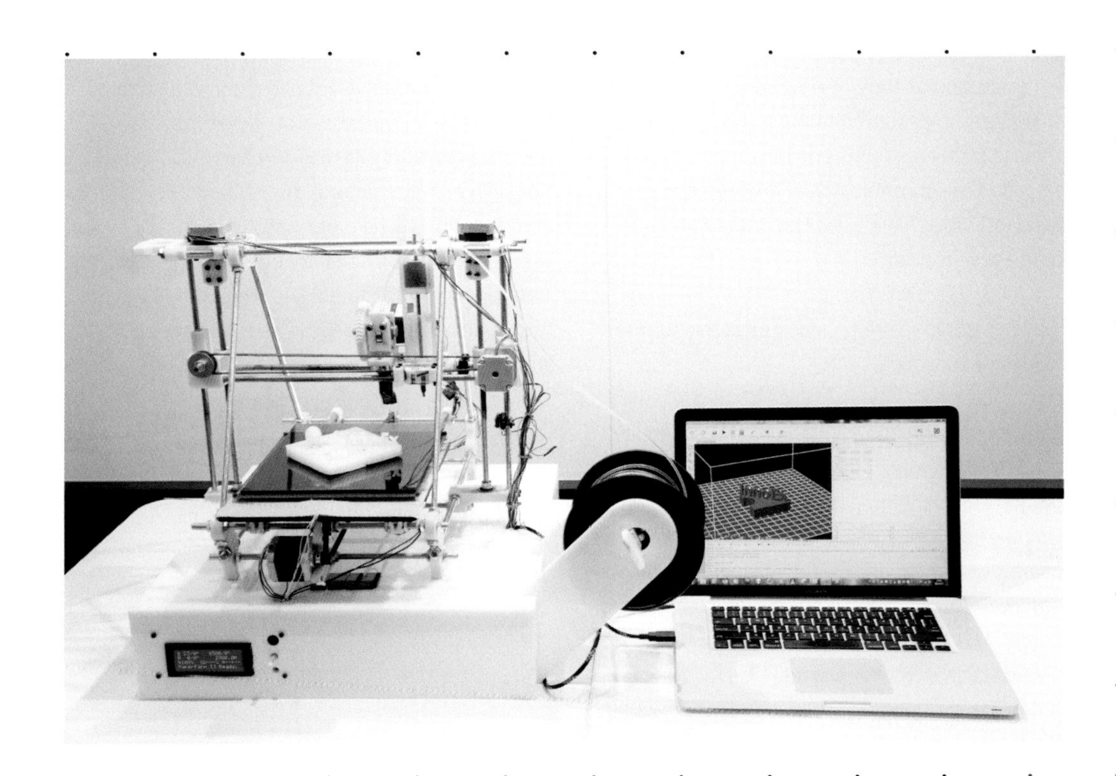

Modified 3D printer for printing electrical conductors and circuits. The project was part of a 2013 cooperation between the University of St. Gallen and Stanford University.

PROTOTYPING AND STORYTELLING

VIDEO PROTOTYPING

Video prototyping is a technique with a history that goes back to Apple in the 1980s. It is increasingly being used across all phases of design thinking. Using this technique, videos of complex scenarios can be developed, recorded, and shown with relatively little effort. Usage is not limited to software interfaces (as with Apple in the beginning), but can be used for services, processes and business model prototypes. Modern computer programs such as iMovie and, increasingly, advanced reflex cameras with video functionalities enable even beginners to quickly develop good videos. Additionally, many video-editing programs can now be directly used on smartphones.

Another development is the building of virtual video prototypes. Web portals such as goanimate.com enable the easy generation of comic-based video prototypes. The producer can choose between numerous characters, set up a variety of scenes, and even choose the voices of characters. With a little effort to learn the ropes, design thinking teams can make videos within a few hours without professional outside help. These animated prototypes can be used in every project phase, however their main value lies in the early stages of design thinking projects, as the costs are low and little development effort is required.

Another popular form of production for video prototypes in design thinking are stop-motion videos. For these, hundreds of pictures of a given situation are chronologically ordered and compiled in a video. This gives the impression of a time-lapse movie.

Video prototypes are typically effective for the following purposes:

Explaining and testing: Video prototypes are a good medium for conveying complex stories for testing and explaining prototypes in an easily transportable way. A video can be presented on a computer, shared on social networks, or uploaded onto well-known video platforms such as Vimeo.

Visualizing the future: Another use for video prototypes is explaining cutting-edge future scenarios. Films such as *Preview of the World from Tomorrow* (Kerstan & Schmuckler, 1972) from ZDF are interesting examples. The use of these videos generally makes most sense in the dark horse phase of design thinking, since this phase is about generating prototypes which lie further in the future.

Despite the numerous possibilities of video prototypes, they also have disadvantages. On the one hand, very well produced videos can make the impression that the prototype presented already exists, and this may awaken false expectations. This effect can greatly influence the quality of feedback. On the other hand, the danger exists that the audience does not evaluate the prototype, but the video itself. They then judge the prototype as being better or worse than they otherwise might have.

TOOLS

- GoAnimate.com
- iMovie — Video editing program from Apple

TIPS AND TRICKS

- Start early with video prototypes.
- Pay attention to permissions to film.
- It is preferable to produce more video material than you need so that you have editing options.

Video prototype to demonstrate a service concept for collaboration between tax consultants and mid-sized businesses

PROTOTYPING AND STORYTELLING

SERVICE BLUEPRINTING

With the help of service blueprinting, service prototypes can be systematically structured and described in a uniform way. A service blueprint comprises various components:

Physical reference point: A physical reference point for a patient's activity in a hospital is, for example, admission. The physical reference point often provides information and evidence on events within the service.

User activities: User activity components refer to all the behavior of a user within the context of a process or service. Returning to the hospital example, this could be filling out a medical history and general information form at reception.

Frontstage activities: The frontstage activities are carried out by the service provider and generally directly relate to user activities. In the hospital example, this might be a doctor's examination of a patient.

Backstage activities: Backstage activities have a direct causal relationship with the services provided, but the service provider carries them out the background, invisible to the user. In the hospital case, this is the invoicing of the treatment. Patients are seldom in direct contact with these processes, since they take place between the hospital and insurance companies.

Supporting activities and systems: This covers all the steps relevant, but not directly connected, to the case. In the hospital, this could involve the sterilization of examination and theater materials, but also the computer systems used to process patient and treatment data.

The particular advantage of blueprinting is the representation of complex situations using a comparatively simple technique. These representations are appropriate when making the first feasibility tests with experts, and provoking new ideas in testers.

MATERIALS

- Whiteboard or *Microsoft PowerPoint* presentation
- Markers
- Index cards or *Post-its*
- Images for visualizing concrete situations

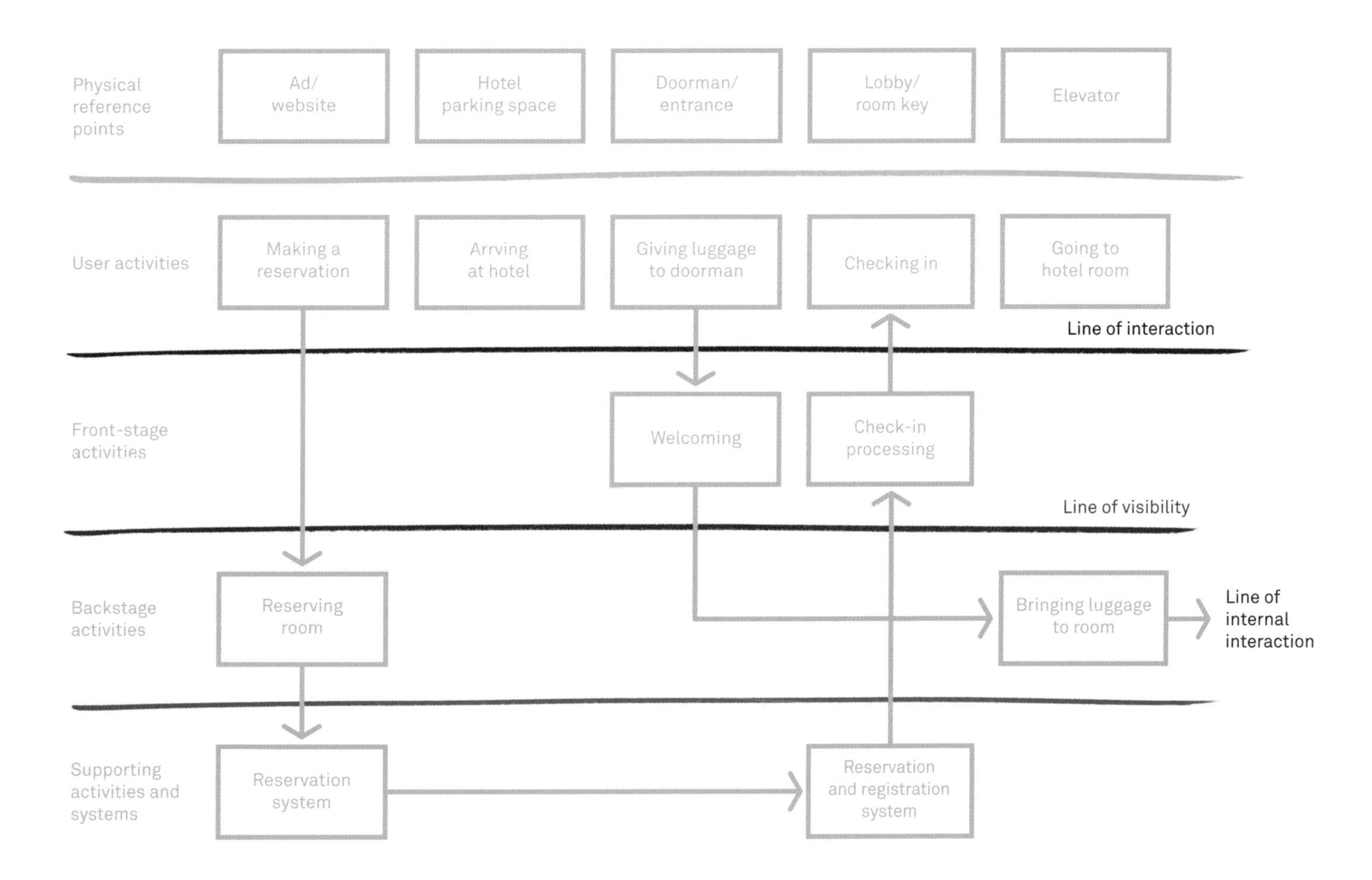
Physical reference points
Ad/ website
Hotel parking space
Doorman/ entrance
Lobby/ room key
Elevator
User activities
Making a reservation
Arrving at hotel
Giving luggage to doorman
Checking in
Going to hotel room
Line of interaction
Front-stage activities
Welcoming
Check-in processing
Line of visibility
Backstage activities
Reserving room
Bringing luggage to room
Line of internal interaction
Supporting activities and systems
Reservation system
Reservation and registration system

PROTOTYPING AND STORYTELLING

SKETCHES AND SCRIBBLES

Sketches and scribbles are hand-drawn representations of products, services or business model scenarios. These drawings are often made in the first phases of design thinking, in which speed is important and prototypes are of low resolution. Scribbles and sketches are created with simple materials and do not require any prior experience. A talent for drawing is helpful in some situations, but our experience has shown that everyone can draw well enough.

Sketches and scribbles show either just the product in focus or important components, but they can also already contain further information for use or embedding in an extended business model context. The combination of multiple scribbles can also tell a story.

Drawings can be useful when the design thinking team is traveling and has no other material for building prototypes available. A marker and a piece of paper are often easily accessible, so ideas can at least be roughly put down.

Sources

Buxton, B., "Sketching User Experiences — Getting the Design Right and the Right Design", 2007

TIPS AND TRICKS

The only tip is: Everyone can draw! Overcome your inhibitions and draw.

MATERIALS

- Pencils and erasers
- Paper
- Index cards or *Post-its*
- Images to visualize concrete situations

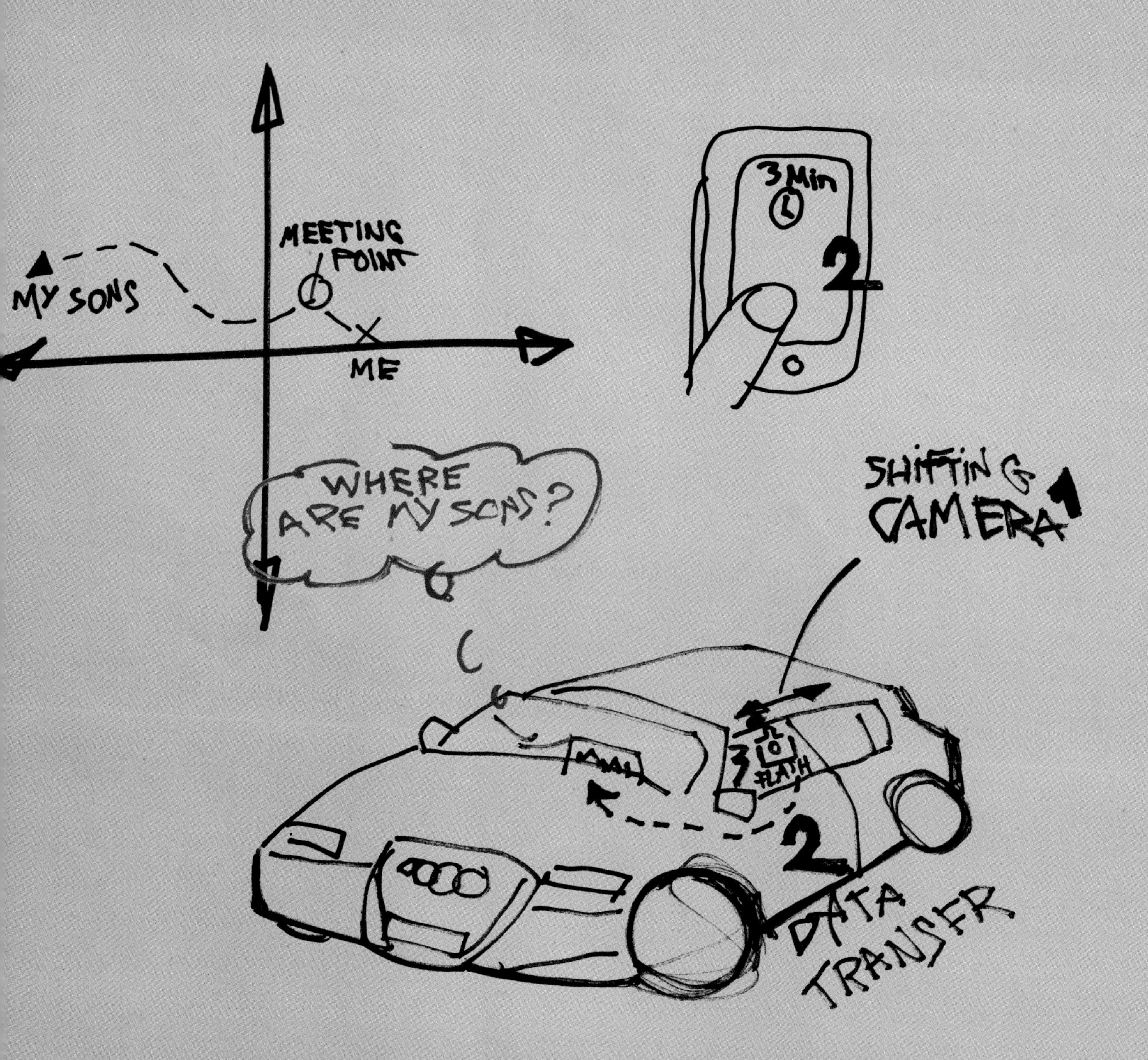
MEETING POINT
MY SONS
ME
3 Min
2
WHERE ARE MY SONS?
SHIFTING CAMERA 1
FLASH
2
DATA TRANSFR

PROTOTYPING AND STORYTELLING

PHOTOSHOP PROTOTYPING

Photoshop prototyping is another form of low-resolution prototyping. By using photos and images of specific situations to create montages, scenarios of product use or service concepts can be demonstrated. For an automotive manufacturer, for example, we prepared an outdoor parking space concept. To show how a display showing parking space status might look — that is, whether a parking space is either occupied, available or a car is suboptimally occupying the space — we laid boxes over the occupied, free and poorly-used parking spaces using photo-editing software. By using manipulated photos, we were able to show the basic principle behind our ideas without extensive text.

The advantages lie in the speed and simplicity of the prototypes. At the same time, testers easily understand these prototypes.

TIPS AND TRICKS

Learning to use photo-editing tools requires lots of time. Beginners can therefore also use *Microsoft PowerPoint* to more easily manipulate photos.

TOOLS

- Still camera
- Photo-editing software: *Adobe Photoshop/Photoshop Elements* or an open-source alternative like gimp.org
- Printer

Prototype for a parking detector in cities and outside parking garages (low-resolution representation); project at the University of St. Gallen (2012)

COMBINED PROTOTYPING

Combined prototypes are on the border between low-resolution and high-resolution prototypes. They are generated using paper and other materials, as well as prototypes made of paper and other technologies. For example, when testing remote support for the overhaul of a laboratory machine, we used a mobile phone with Google Hangouts. For these prototypes, we fastened a mobile phone, with the camera facing forward, to the tester's head using zip ties. Then, using the mobile phone, we initiated a Google Hangouts (video-telephony) session to bring a remotely-contacted expert into the situation. A headset connected to the mobile phone ensured that the tester could receive all instructions via headphones. Using such prototypes, multiple techniques and technologies can immediately be put to use and a relatively complex scenario can be tested in only a few hours.

Speed is the advantage of using combined prototypes. Combined prototypes are higher-resolution prototypes than prototypes made from only one material, they are therefore better able to be experienced and tested.

TOOLS

- Open hardware
- Materials needed for the specific procedure

NOTES

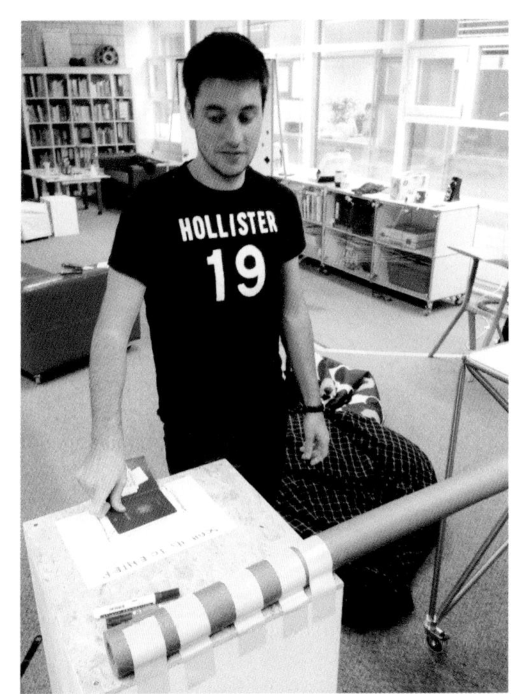

1

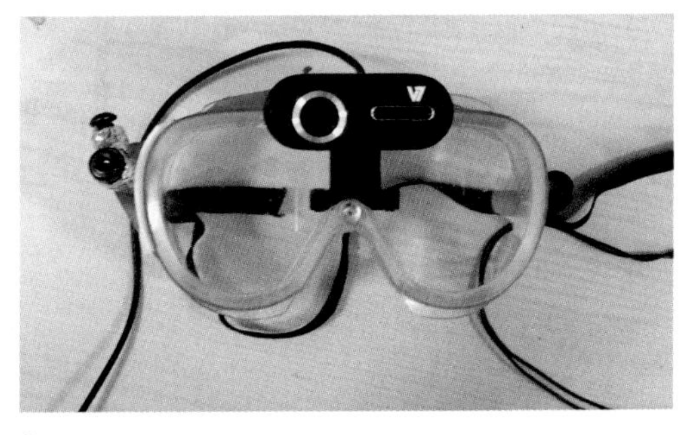

2

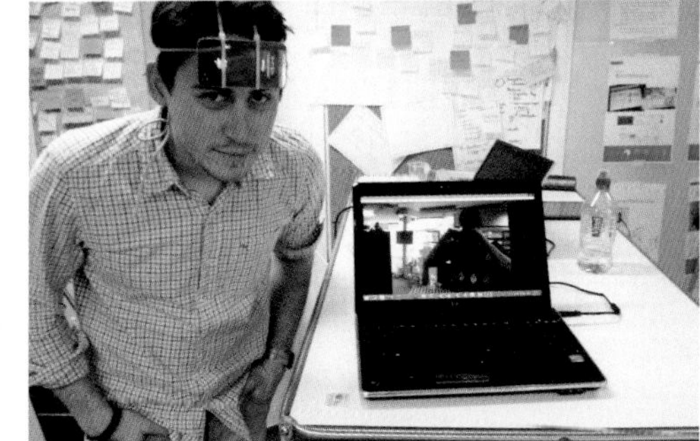

3

4

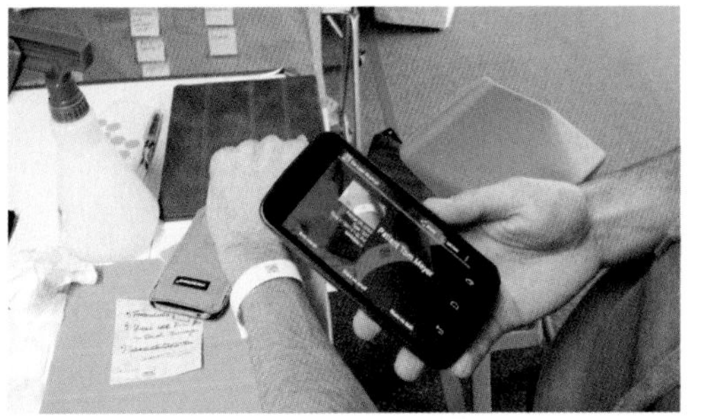

5

1 Prototype for quick and easy registration using a passport or personal ID card, University of St. Gallen and Pontificia Universidad Javeriana Cali (2012)
2 Prototype of a helmet camera to aid technical support providers during maintenance work, Stanford University and University of St. Gallen (2014)
3 Prototype to simulate the remote support of machines, University of St.Gallen (2014)
4 Integrated prototype of a laboratory machine (2014)
5 Prototype of an armband for registering patients in hospitals, University of St. Gallen (2014)

PROTOTYPING AND STORYTELLING

TOWN PLANNING

Town planning is a category of prototypes that present architectural or building-related aspects of a solution. This kind of prototype is also useful when developing services. We used this method to develop a new consulting concept for asset management. An element of the concept was also the reworking of the consulting spaces within the bank.

For high-resolution prototypes, three-dimensional CAD drawings come into consideration. Three-dimensional drawings tend to demonstrate what a new scenario might look like at a glance. However, low-resolution prototypes, such as mock-ups of rooms or models for city planning, can also be represented using paper, cardboard, wood and other materials. If the low-resolution prototypes are built up in modules, the tester can show his or her own ideas and imaginings.

With the help of a city planning model, we presented city center receiving and repair stations for cars connected to a network of city outskirts workshops for a company in the automobile industry. The receiving stations were externally inspired by the design of Apple products, and were shown as prototypes.

Town planning prototypes usually integrate testers and observers very quickly and provoke both good and bad emotions.

TIPS AND TRICKS

FabLabs offer possibilities to quickly build small architectural prototypes.

TOOLS

- Paper, wood, cardboard, glue, etc.
- 3D drawing tools, e.g. Google SketchUp

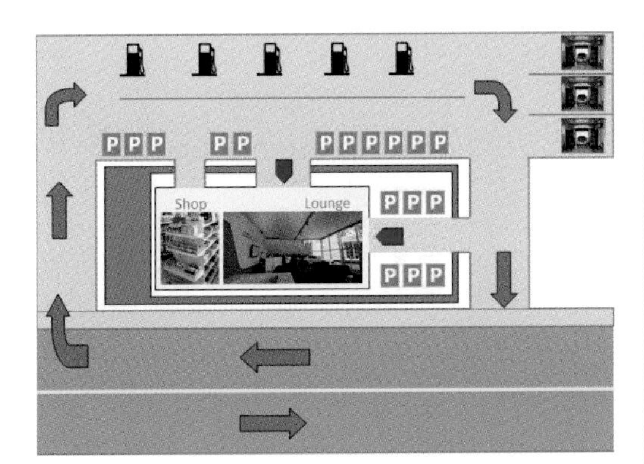

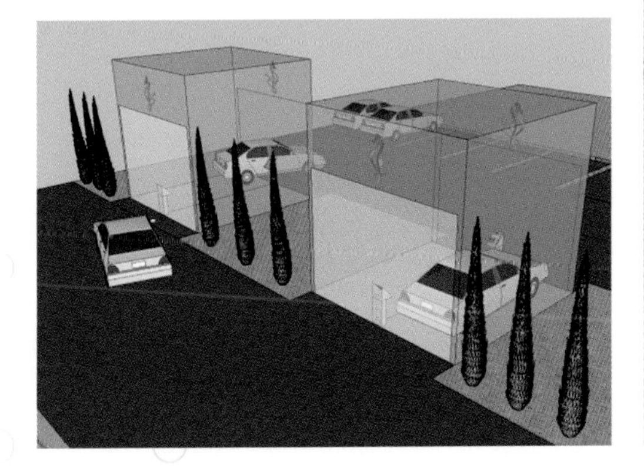

Concept for a car repair receiving station in large cities following the example of the Apple Store, University of St. Gallen/Karlsruher Institute of Technology (2014)

PROTOTYPING AND STORYTELLING

BUSINESS MODEL PROTOTYPES

Business model prototypes show the commercial connections between various actors such as customers, producers, suppliers, and so on. The goal is to systematically demonstrate the business relationships between, and motivations of, the stakeholders involved. The best-known framework for describing business models is the Business Model Canvas, by Osterwalder, Pigneur and Clark (2010). This model structures business models along nine dimensions: customer segments, customer relationships, channels, value propositions (of products and services), key resources, key activities, key partners and cost structure and revenue streams. The model is excellent for establishing transparency, and analyzing complex business models.

Nevertheless, in representations for testing ideas the model usually proves itself too "technocratic". By this we mean that the tester and user often struggle to imagine themselves within the parameters of the dimensions the framework describes, and often do not have the time to bring in their own sense of fantasy.

For this reason, besides the canvas, context-specific business model prototypes are built in design thinking projects. They show business models using either as little as just sketches and scribbles or use the aid of other materials and elements of the business model as well. The advantage is that, in contrast to the pre-defined categories, these can be individually gone into with the particular business or customer context in mind.

Business model prototypes can be used during all the phases of design thinking. However, we try to place consideration of business concerns, which this prototype category prioritizes, behind the identification of real user needs.

Sources

Osterwalder et al., "Business Model Generation: A Handbook for Visionaries, Game Changers, and Challengers", 2010 (1st ed.)

TOOLS

- Business Model Canvas
- Paper, cardboard, markers, etc.

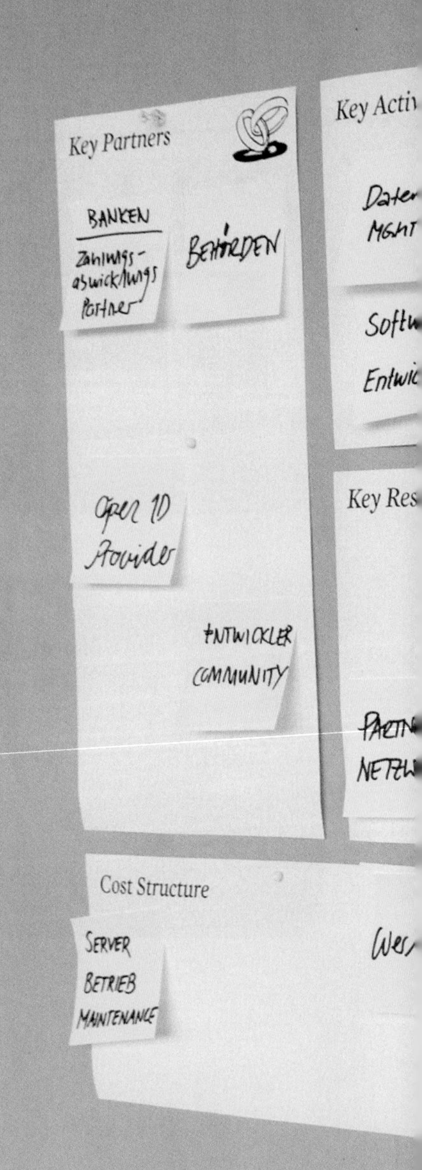

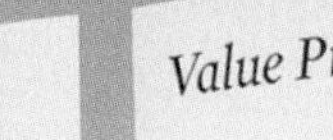
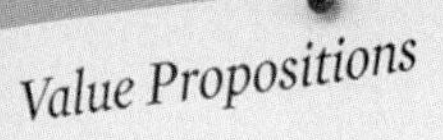

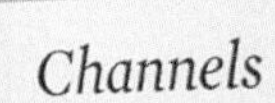

Value Propositions
Faster
Instant
Administration
on the go
Trusted
Business
Platform
the small
business
community
Customer Relationships
TRUST
Channels
WEB-
PLATTFORM
WOM
Support
Customer Segments
Klein-
Unternehmen
Bis 50 Pers.
BEHÖRDEN
Service
Anbieter
MASS
MARKET
Revenue Streams
SCHULUNG
CONSULTING
Premium-
Angebote

PROTOTYPING AND STORYTELLING

CONFLUENCE DYNAGRAM

By Martin Eppler and Sebastian Kernbach

The confluence dynagram is a multidimensional configurator in the form of a radar diagram. It can be dynamically inscribed and adapted in order to capture and make the capabilities of prototypes visible, including interdependencies and potential limitations.

In addition, the confluence dynagram enables design thinking teams to see and discuss the interdependencies between various factors and their effects on the profile of the prototype. This enables the team to select a profile of factors when preparing a prototype. The learning from the prototype test can, in turn, be integrated into the confluence dynagram, in which the previously assumed interdependencies and assumed profile can be changed. By saving different statuses of the confluence dynagram, the prototype development is documented and can be later reflected upon.

Sources

Eppler, M.J., Kernbach, S., Wiederkehr, B., & Gassner, P., "The Confluence Diagram: Embedding Knowledge in Interaction Constraints", 2014

TOOLS

- www.dynagrams.org

TIPS AND TRICKS

Identify the factors: List the factors that comprise the prototype and select a minimum and maximum value for each factor.

Group factors and define dimensions: Group the factors that belong together in a given dimension and name the dimension.

Establish interdependencies: Discuss the interdependencies between the different factors in a dimension, but also the factors that bridge multiple dimensions. You can define each of these interdependencies as a "more leads to more" or "more leads to less" interdependency. This is not about numerical exactness, but tendencies. The higher the number of factors that a factor influences, the bigger the circumference around the factor's sliding knob.

Establish a profile for the prototype: By sliding the knobs, the interdependencies can be made visible. The team can move different knobs, thereby establishing a profile for discussion.

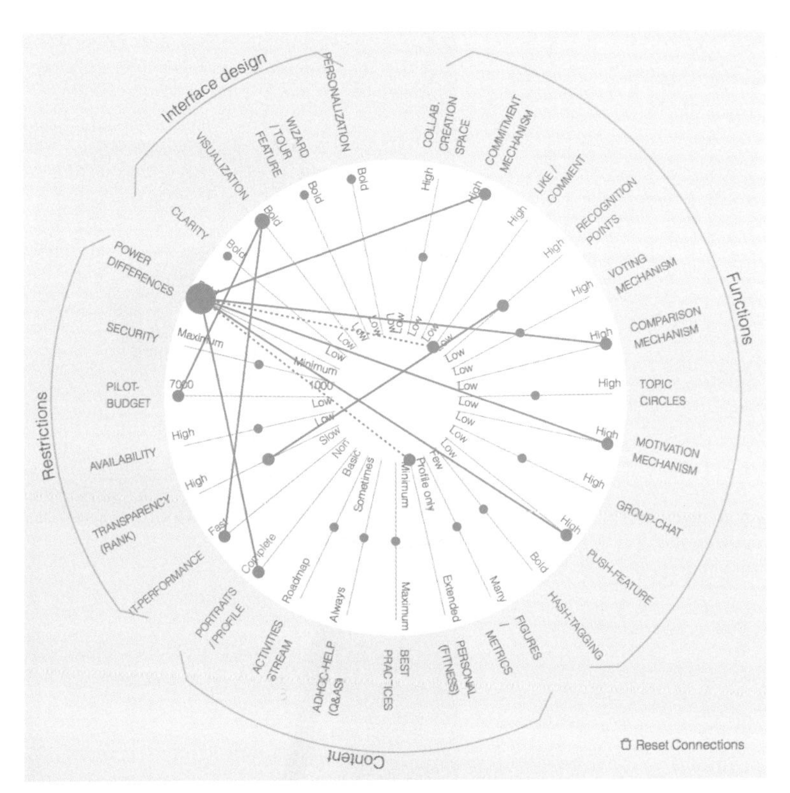

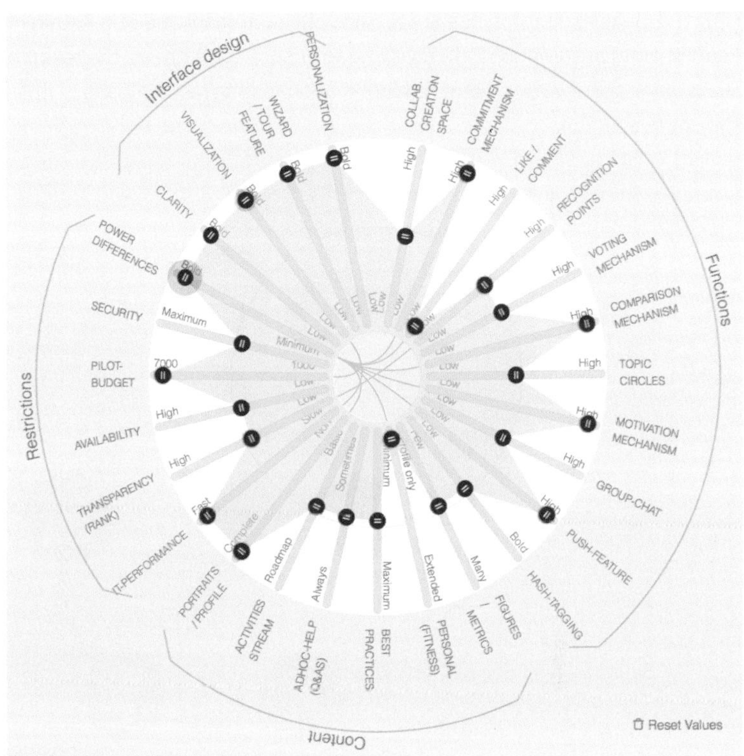

Confluence dynagram of a design thinking prototype for an insurance company (2015)

TESTING

Now the ideas and prototypes are tested with customers and other stakeholders. Learning and assumptions are reviewed to ensure correctness, the result being that they are either right or wrong. Team reflection upon testing results leads to an improvement of the entire solution.

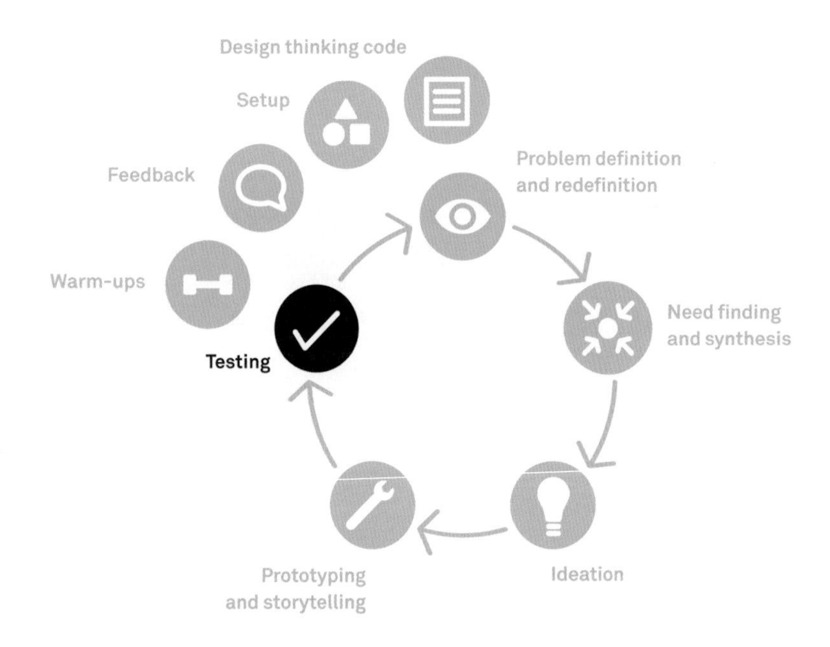

"Prototype as if you know you are right, but test as if you know you are wrong."
(D.School Stanford University, 2015b)

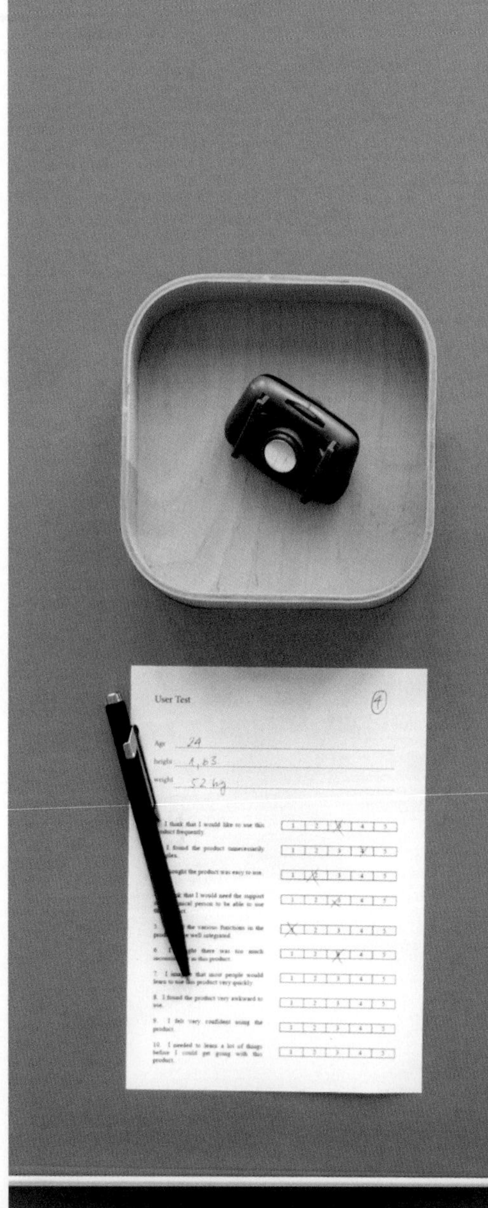

TESTING

CONSUMER CLINICS

When using consumer clinics as a testing approach, customers and users are presented products and services in actual use. Customers and users can then express their opinions. In addition to new products and services, existing products and services can be shown as a point of comparison. The concept was developed in the automotive industry in the 1960s with the goal of eliciting customers' opinions on car models.

For consumer clinics, the people invited should be representative of the intended customer and user group of the product or service being presented. Consumer clinics are typically carried out in three phases: preparation, demonstration and follow up. In each phase, testers fill out a survey that asks about the users' opinions of the particular product or service. This data is then available for extensive evaluation and analysis.

The number of testers can range from 10 to several hundred. In contrast to testing within focus groups, where the products and services introduced are still in the formation phase, the products and services in consumer clinics are already actually usable. In this sense, consumer clinics not only show the personal opinion of the tester on the product demonstrated, but also show its application and use. Through this, consumer clinics offer a good method of testing high-resolution prototypes, since they already offer an (almost) complete simulation of processes and functionalities. In the early phases of a design thinking project, this technique is not as highly recommended.

TIPS AND TRICKS

- Select the test group and individual testers.
- Preparation: Distribute the survey on what the product testers hope to get from the product, what their expectations are, etc.
- Implementation: Demonstrate the product or service and try out the application or use with the tester.
- During the demonstration phase, observe and document the tester's reaction.
- Follow up: Using a survey, the tester is again asked to share his or her opinion and overall impression of the product tested. Surveys also ask whether the product fulfilled the tester's expectations, what he or she liked and why, as well as additional possible personal uses, etc.

TESTING

USABILITY TESTING

The techniques and procedures in which testers use a product or service under set uniform framing conditions are discussed under the term usability testing. With this method of testing, the tester must carry out a specific, pre-determined task using the product or service.

For example, when testing an online shopping portal, testers were given the task of purchasing specific goods, or changing their profile information in a customer portal. During tests, all movements, steps and actions were observed and noted using cameras or pre-installed software products.

There are different variations of usability testing available:

Cognitive run: The focus of this testing form is the tester's cognitive processing of the product or service. Particularly in the case of software, split-seconds can decide whether a product will be perceived as making sense or not. If, when a user stands in front of a ticket machine, the number of buttons or entry fields they must search through is too detailed and cumbersome, this would be discovered.

Activity analysis: This explores the tester's general behavior when using the product or service. With formal analysis, technological tools will show and protocol the whole flow of activities. With informal analysis, activities are simply run through and an assessment of the prototype's strengths and weaknesses is made at the end.

Thinking out loud: Thinking out loud sounds almost primitive. The technique is not demanding, but it is effective and efficient! Using the technique, testers express their thoughts, reactions, and what they want to do next aloud when using a system. Through this method, a co-present observer can understand and write down the tester's thought processes.

Survey: Surveys, a traditional instrument, can bring numerous aspects of usability to light, for example in computer systems. This technique is an indirect method of usability testing, as it does not directly draw upon or involve the product or service being tested. Both qualitative and quantitative questions are considered for inclusion in the survey.

TIPS AND TRICKS

There are numerous service providers in the activity analysis field that carry out professional usability testing, for example www.userfeedback.ch.

TOOLS

- Prototypes from the project
- Notebook and pens/pencils
- Video camera to record testing

TESTING

NABC PITCH

The NABC pitch is a short, to-the-point description of a new idea. It's now also commonly known as an elevator pitch, since ideally, the description of the project's intention is so brief that one can explain and sell it to his or her boss during an elevator ride.

The abbreviation NABC stands for the keywords need, approach, benefit and competition. The pitch should address all of these keywords.

Need: What needs does the customer have, and which problem should be solved?

Approach: How does the new idea address the needs? How will the problem be solved?

Benefit: Through this concrete idea, what value does the solution provide the customer with?

Competition: Which features differentiate the solution from others in existence?

NABC pitches are an especially helpful tool for the design thinking team when stepping back from the complexity of their own thought processes and formulating the core of a product or service. If one is able to express a solution in 30 to 60 seconds with an NABC pitch, he or she has located the core of the idea.

When testing, the NABC pitch — with or without the prototype — is presented to the tester, and immediate feedback on the idea can be gathered. The advantage of the NABC pitch is that it enables the tester to quickly understand and evaluate the essence of an idea. This method is useful for testing low-resolution prototypes in a dynamic environment, such as the street. A detailed explanation with lots of detail would take too much time, preventing many testers from participating.

TIPS AND TRICKS

- Put together an NABC pitch that describes your idea in a way that is efficient and to the point. Remember that the pitch should include an explanation of the problem, your idea, the value for the customer, and what makes your idea unique.
- Explain your idea to the testers and ask them for brief feedback.
- Consider the following factors when developing your pitch:
  - What problem are you solving with the prototype?
  - What is the prototype, and how does it solve the problem?
  - How does the customer get the prototype?

TOOLS

- A head for thinking with!
- Pens/pencils and paper

TESTING
PECHAKUCHA

PechaKucha is a newer form of the NABC pitch. Here, the goal is also to present a complex idea within a short period of time, however here, the presenters use slides and have 20 x 20 seconds to present exactly 20 slides. The presentation software is often set so that after 20 seconds, the next slide appears. Because of this, the presenter must be short and to the point. This form of presentation does not allow for extensive text on slides — the focus should be on images. This increases the presentation's clarity, which makes it easier for the testers to understand.

There are now many PechaKucha events worldwide. Design thinking teams can also share their presentations and have them evaluated.

As a testing approach, PechaKucha is similar to NABC pitches in that the participants are presented with the idea and feedback can be collected directly.

Sources

PechaKucha, 2015; from www.pechakucha.org

TIPS AND TRICKS

- If possible, use only images for the slides; audiences need less time to grasp pictures than text.
- Practice the presentation several times, and do not forget to activate the automatic slide-changing function.

TOOLS

- www.pechakucha.org
- Microsoft *PowerPoint*
- Prezi presentations (www.prezi.com)
- *Keynote*

NOTES

TESTING

NEED-FINDING TECHNIQUES

Some need finding techniques can also be used for testing prototypes with users or in labs. This has the advantage of enabling a combination of the need finding and testing phases, especially in large, lengthy projects.

The following techniques are appropriate for both need finding and testing:

Focus groups: The manageable number of participants make focus groups useful, particularly when considering and discussing low-resolution prototypes either conceptually or with presentation or discussion. For more refined prototypes, consumer clinics are better.

Personas: Personas can also be used as a technique for the initial testing of products and services. Particularly in design thinking labs, personas can help the team members put themselves in the position of the user's character, and thereby carry out an internal "substitute" testing. For this, a team member takes on the role of the selected persona. This team member then maintains and acts out the characteristics of the persona, and tests the prototype. Using this technique, problems can be identified early and solved, in some circumstances even before they are communicated to the "real" tester.

Interviews: Interviews are also an excellent opportunity to test existing prototypes. Make sure that the test is carried out after the actual interview takes place, otherwise you run the risk of discussing only the prototype.

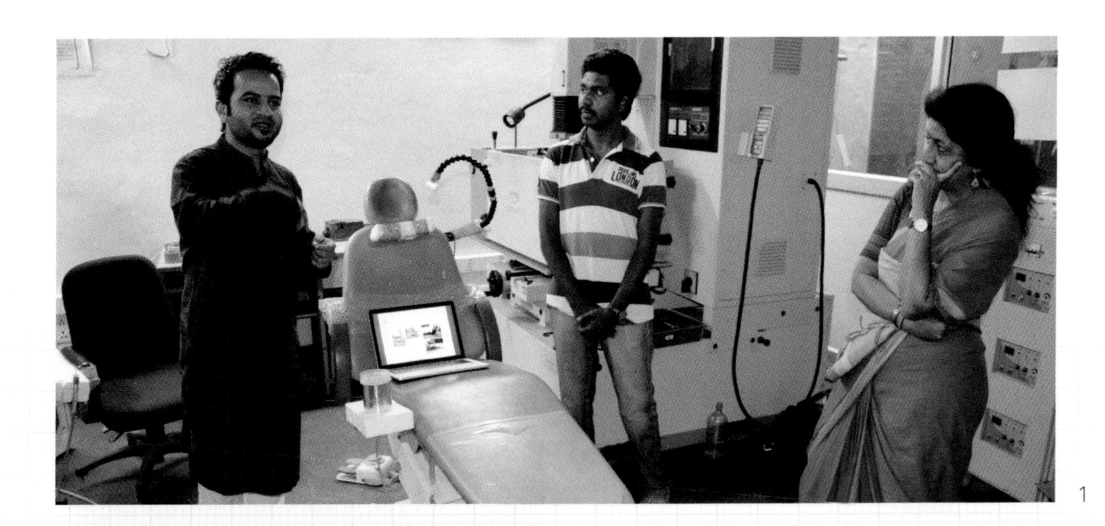

1 Testing a new chair for use in a dentist's surgery, Kanpur, India (2015)

2 Testing at the end of a focus group, Shanghai, China (2015)

WARM-UPS

Warm-ups are little exercises that prepare the body and mind for a challenging task — just as in sports, music and the performing arts.

In design thinking, it is also helpful to warm up from time to time. Brainstorming from a standing start can be difficult and laborious. Participants are usually not immediately in the correct, creative and inspired frame of mind. In these cases, warm-ups stimulate thinking and help participants concentrate on the task to come.

In principle, warm-ups can be used in every design thinking phase, particularly when work appears arduous and stagnant. In this chapter we describe several warm-ups, but there are, of course, countless more.

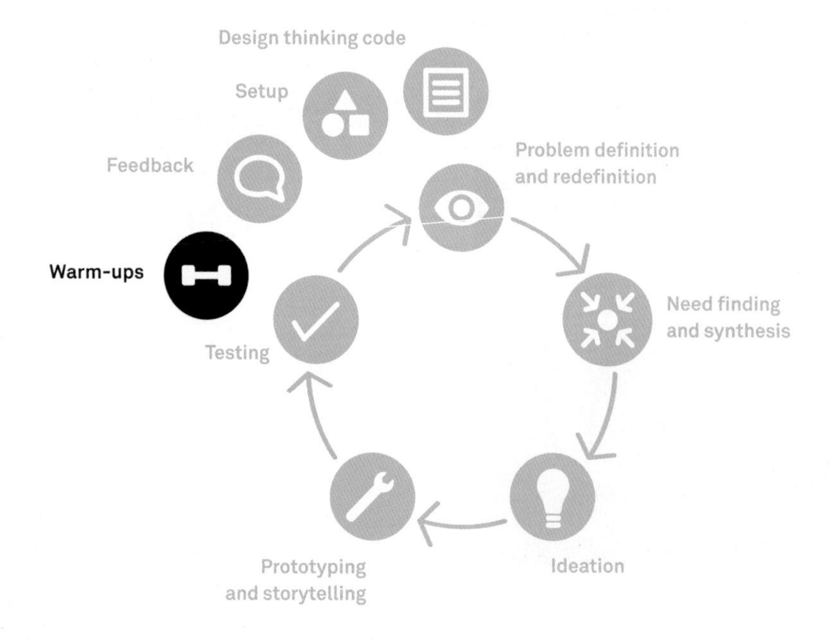

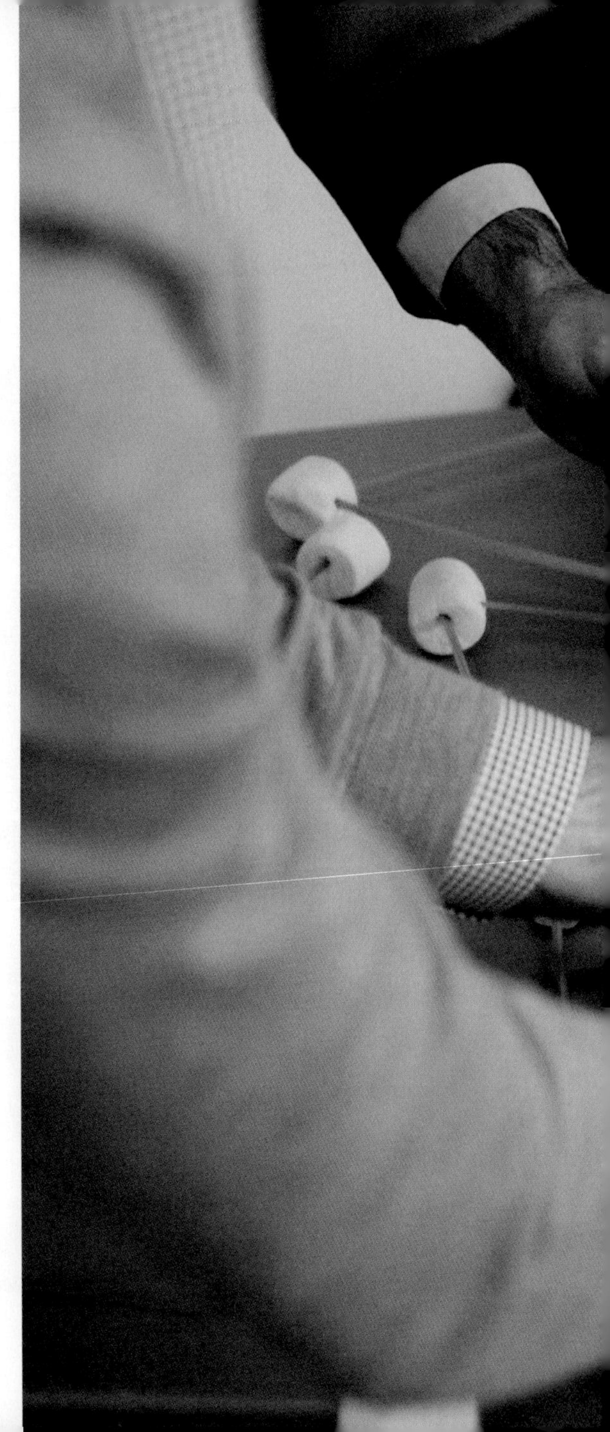

WARM-UPS

SPAGHETTI TOWER

The spaghetti tower is a fun and entertaining exercise that motivates participants to build and experiment. The task is to build the highest possible freestanding marshmallow and spaghetti tower within 10 minutes. Additional materials are not allowed.

This exercise can also be more than a warm-up. Teams can also use the spaghetti tower to assess and evaluate a team's approach to planning and execution. The group discussion that follows can inspire reflection.

Variations are also possible:

Repetition: If one wants to show that copying is also part of the innovation process, then a second run-through of the spaghetti tower can be carried out after the first round. Between the two rounds, the teams present their tower and explain their approaches to construction. This new knowledge and the team's own experience flow into the second round, when they have the opportunity to build their learning into second towers. Following the motto "fail forward" within the scope of prototyping, this variation of the exercise helps make false assumptions visible early in the innovation process.

Paper tower: Instead of spaghetti, newspapers and tape can be used. This variation is somewhat cleaner (but not as much fun).

TIPS AND TRICKS

- Procure sufficient spaghetti and marshmallows.
- Divide the group into teams of 3–4 people, and give each team the same amount of spaghetti and marshmallows.
- Give the teams 10 minutes to build the highest tower.
- After 10 minutes, stop building and compare heights.

MATERIALS

For the spaghetti tower

- Spaghetti
- Marshmallows
- Garbage bags (for cleaning up)

For the paper tower

- As many newspapers and magazines as possible
- Tape

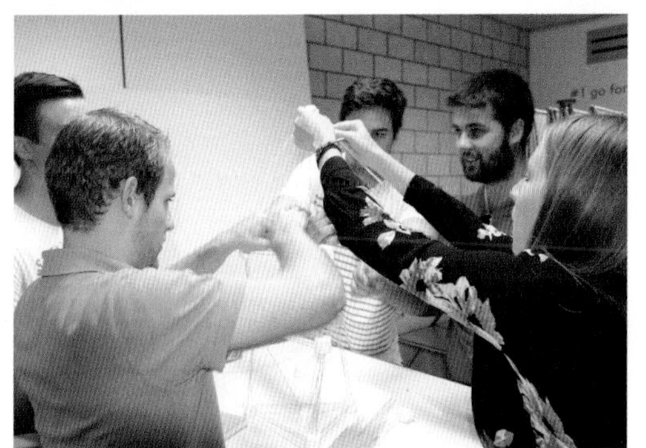

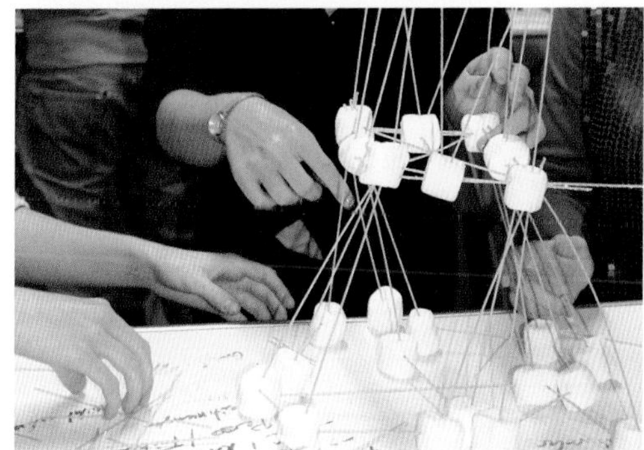

Spaghetti tower warm-up at the University of St. Gallen (2014)

WARM-UPS

YES BUT, YES AND

Who hasn't heard the phrase "Yes, but...."? It is used day in and day out in workshops, conversations, and brainstorming sessions. The effect is usually that people do not truly collaborate and build upon each other's ideas, but instead build up parallel worlds of ideas and thoughts in separation from one another. The "yes but, yes and" technique takes a playful approach to give this transparency.

TIPS AND TRICKS

Round 1: The group is divided into subgroups of two participants. Each subgroup receives the task to plan a trip from Zurich to Shanghai, for example. Each group member is instructed to bring their ideas into the round, but only with the preface "yes, but...". Give the groups two minutes to plan the trip. At the end of the task, ask them where the trip ended up.

Round 2: Retain the same group setup and assign the same task, the only difference being that participants may contribute using only the preface "yes, and...". After 2 minutes, ask where the trip led.

Typically, when the teams use "yes, but...", their trip does not depart. In contrast, the ideas flow when using "yes, and...". With team members, reflect upon why and which processes made this the case.

This warm-up is particularly appropriate for ideation phases.

NOTES

Yes, but...
Yes, and...

WARM-UPS

RACES

Races is a warm-up that brings momentum to tired teams. Groups of 4–5 participants are given the task of covering 10–15 meters with only four feet and six hands touching the floor. The team that covers that distance the fastest wins. Teams are given five minutes of preparation time to practice for the race.

The high athletic exertion quickly reactivates tired teams and prepares them for a task requiring concentration and action. The disadvantage is that this exercise is not possible with every group of people, and some environments do not offer sufficient space.

TIPS AND TRICKS

Lots of water and chocolate should be readily available!

TOOLS

- Start and finish lines
- A large hall or space
- Chocolate, cookies, etc.

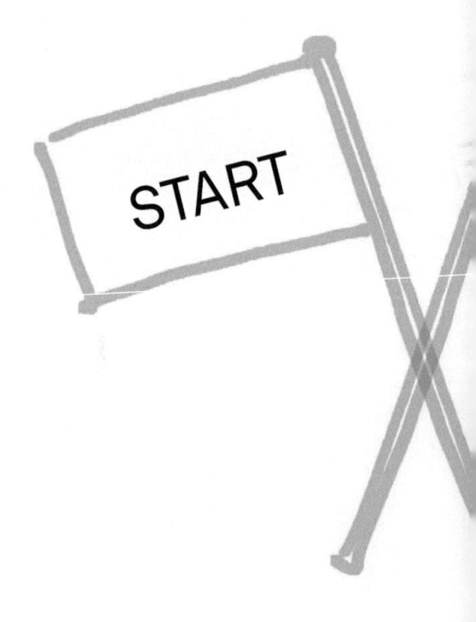

Race at a design thinking bootcamp at the University of St. Gallen (MBI Master's and MBA program) in rooms provided by the Swisscom BrainGym in Bern. The BrainGym offers Swisscom employees ideal spaces to use human-centered design methods (2014).

WARM-UPS

ASSEMBLY

Assembly takes 5–10 minutes, and helps team members to get to know each other better in a playful manner. Over the course of the exercise, the group is asked a range of questions and must order themselves into a pattern of their own finding. For example, the group may organize itself according to the participants' shoe sizes. Another task would be to organize themselves according to favorite vacation spots. As soon as the group has determined their order, the moderator may ask the participants about their results.

The assembly warm-up works with both small and large groups of up to 40–50 participants. Our experience has shown that it works well at the beginning of workshops or similar events.

TIPS AND TRICKS

This warm-up works best when you orient the questions around the team's actual issues. If the team is, for example, working on a question about clubs and associations, you might ask the participants to organize themselves according to the total number of club memberships they have had. This develops empathy for the topic and is always entertaining.

TOOLS

- A large, clear space

Assembly warm-up at an international executive workshop at the University of St. Gallen (2014)

WARM-UPS

STICK FIGURES

Everyone can draw! The stick figure warm-up is an exercise that encourages participants' ability to draw little graphics. The participants receive *Post-its* and pens/pencils. The task is to draw stick figures showing different emotions, such as curiosity, meanness, criticism, boredom, and so forth. Participants typically get five minutes to complete this exercise. After five minutes, participants bring their *Post-its* with stick figures to the board, organize them into columns according to emotion, and briefly explain them.

This warm-up exercise is particularly useful as a start for brainstorming or prototyping. Through this exercise, the participants experience that despite having possibly not drawn for years, they still have the ability to draw little descriptive pictures.

TIPS AND TRICKS

- Distribute *Post-its* and pens/pencils to all group members.
- Ask the participants to each draw a particular emotion on a *Post-it*.
- Give the participants 5–10 minutes.
- Have the participants stick their *Post-its* on the wall and explain them.

MATERIALS

- *Post-its*
- Pens/pencils
- Stopwatch
- Gong

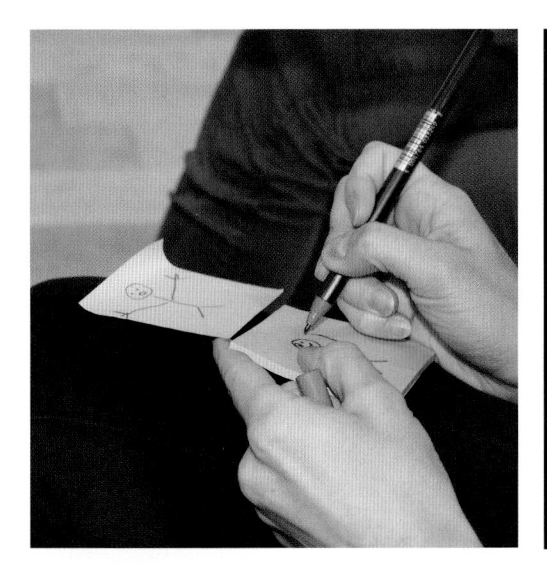

Stick figure warm-up at a company workshop (2015)

FEEDBACK

Open communication and, in particular, an intensive feedback culture should be part of every design thinking project. Well-structured and constructive feedback helps the design thinking team to identify misassumptions or acquire new knowledge through reflection upon questions and advice.

Essentially, the following rules apply to healthy feedback cultures in design thinking projects:

Constructive: Feedback should be constructive and build knowledge. A destructive feedback culture can quickly lead to frustration and demotivation within the team.

Openness: There are no "dumb" questions or remarks. If a team member has questions or commentary, then these are to be listened to and discussed within the team.

Always: Feedback can be given at any time in the process, not only during the time planned for it.

Practice has shown that team members give each other feedback after certain intervals. Through this, problems on a content or personal level are addressed and dealt with. However, feedback can also come from external third parties. It is advisable to elicit active feedback from participants in projects' intermediate status presentations. The techniques that follow can be used for this.

Feedback round with students from the University of St. Gallen MBI and MBA programs at the Swisscom BrainGym. The BrainGym offers Swisscom employees an ideal setting to use human-centred design methods.

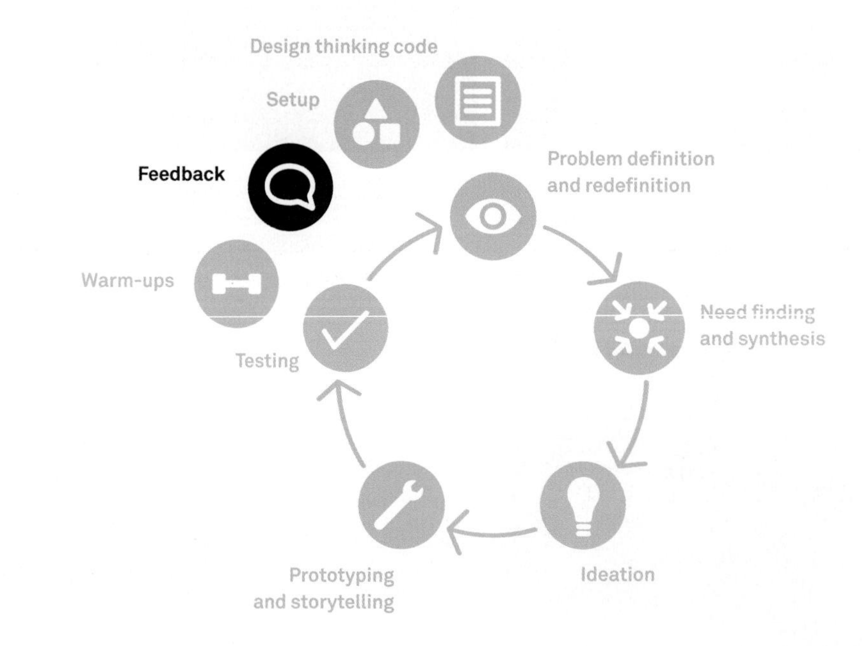

BestPractice
Erfolgsgeschichten
zum mitnehmen

FEEDBACK

I LIKE, I WISH, WHAT IF

The technique "I like, I wish, what if" helps to motivate the open formulation of feedback within teams. The method works excellently in both small and large groups. Practically speaking, for large groups we distribute feedback surveys on every chair so that each participant has a space to record their feedback.

A usage example for a presentation could look like this: "I want product ideas to be interactively demonstrated on-stage through roleplay". "I wish that next time, the audience could be more closely integrated through roleplay". "What if next time, we rehearse the presentation together in advance?"

The "what if" question can also be combined with "I wonder" and "how to". In addition to using the right technique to give feedback, it is also important for team members to listen to each other. Not all feedback has to be immediately reacted to. In reality, the team should decide which feedback should be addressed without delay.

TIPS AND TRICKS

Ensure, particularly with beginners, that the structure "I like" and "I wish" is really adhered to. In our trainer workshops, we actively correct the participants when they do not use it correctly.

TOOLS

- Paper and pens/pencils

NOTES

Team MSB 3000

I like …	I wish …
– critical attitude	– Hang the poster higher to make everything more readable
– open presentation	– speak loudly and clearly
	– optimize time management

FEEDBACK

PLUS OR DELTA

The goal of the plus or delta method is to give a group of people constructive feedback. As in "I like, I wish, what if", the feedback is classified in two categories. The plus category covers the feedback on points that are positively received — that is, that the team should absolutely continue doing or repeat. The delta category includes the points that the team might reconsider or change in future.

The plus or delta technique can be implemented for any type of feedback, regardless of whether it focuses on ideas, activities, prototypes, or anything else.

Sources

Brown, S., Gray, D & Macanufo, J., "Gamestorming: A Playbook for Innovators, Rulebreakers, and Changemakers", 2010

TIPS AND TRICKS

Make sure that the method is used consistently and completely.

TOOLS

- Template cards
- Flipcharts
- Pens/markers

+ quality of speakers
+ interactive exercises
+ big objectives were accomplished
+ opportunity for Q/As were ample
Δ make break-outs cross-functional
Δ go offsite for lunch
Δ look into SlideRocket, Ignite or Prezi for presentation.

FEEDBACK

FEEDBACK CAPTURE GRID

The feedback capture grid is a technique for collecting feedback, but also organizing feedback that has already been collected. The advantage lies in how the feedback capture grid systematizes answers in a way that "first users" also tend to find easy to understand.

TIPS AND TRICKS

The capture grid can simply be drawn on a blank sheet of paper. The page is divided into four sections of equal size. The upper left quadrant is labeled with a plus sign, and contains the positive feedback points. The upper right quadrant is labeled with a delta sign. As in the plus or delta method, the points for reconsideration or change are recorded here. The lower left corner should show open questions, as indicated by a question mark. The lower right corner shows the new ideas prompted by feedback. This corner can be labeled with a light bulb picture.

Like other feedback methods, the Feedback Capture Grid can be used with a variety of group sizes, from small to very large.

TOOLS

- Paper
- Markers/pens

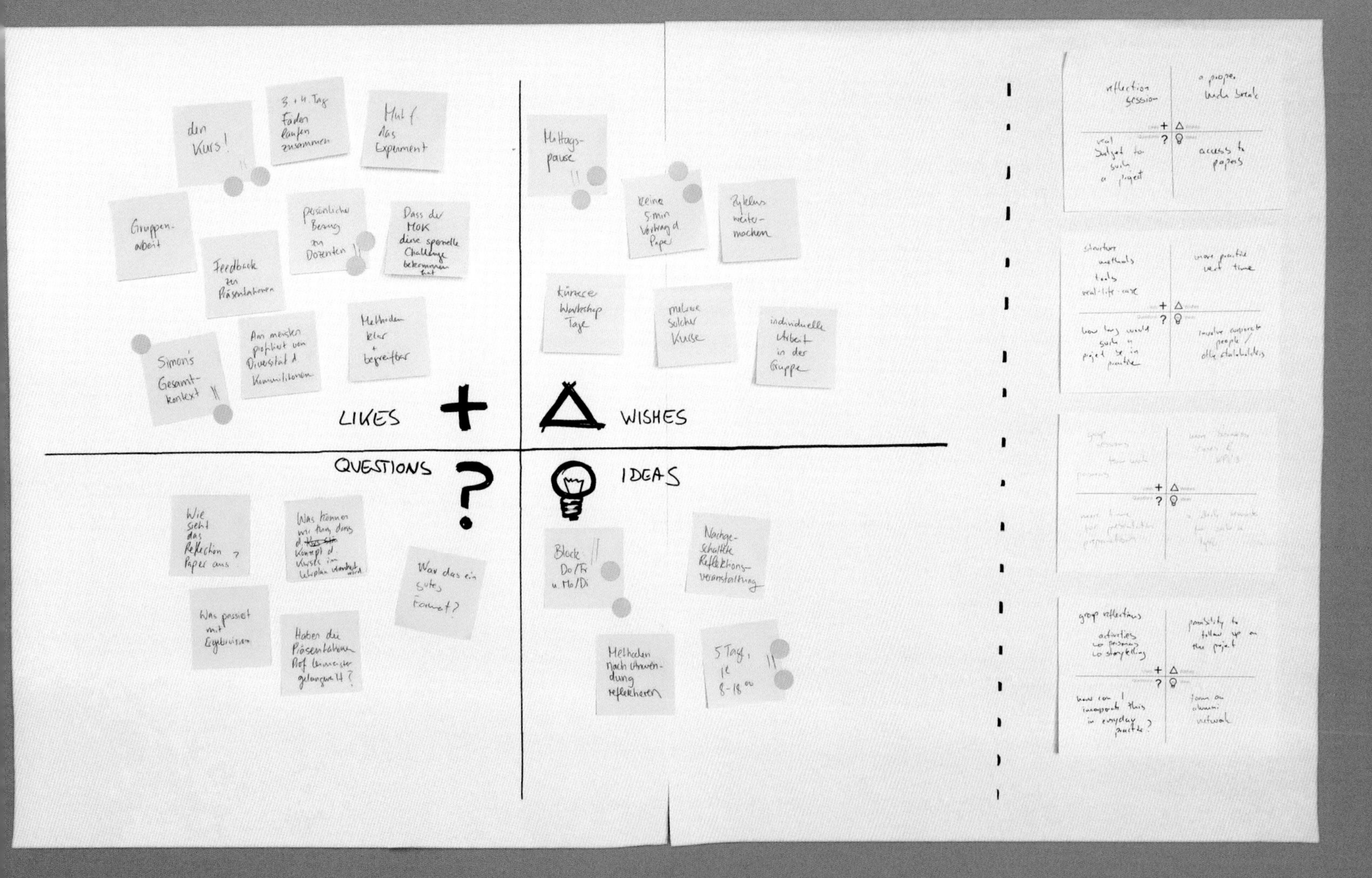
LIKES
WISHES
QUESTIONS
IDEAS
den Kurs!
Gruppen-arbeit
Feedback zu Präsentationen
Mittags-pause
Zyklus weiter-machen
kürzere Workshop Tage
mehrere solcher Kurse
individuelle Arbeit in der Gruppe
Simon's Gesamt-kontext
Wie sieht das Reflection Paper aus?
Was passiert mit Ergebnissen
War das ein gutes Format?
Block Do/Fr u. Mo/Di
Methoden nach Anwendung reflektieren

FEEDBACK

CRITICAL READING CHECKLIST

The critical reading checklist is a technique for giving teams feedback on ideas and perspectives during the design thinking process. The goal is to test validity while assessing whether or not a perspective has potential, is unique, is focused enough or is exciting.

Using the critical reading checklist is easy, and involves posing and answering the following four questions:

1. "What is the point?": Is the idea/perspective customer-oriented? Is it based upon a need? Is it grounded in an insight?

2. "Who says it?", or rather, "How valid is this perspective?": Is the perspective supported by customer and user needs, or can it be be derived from customer or user needs?

3. "What is new?": What is the value of the perspective? For example, was the need articulated via a new approach? Is it positioned within a new user context?

4. "Who does this interest?": How significant is the perspective? Is the team enthusiastic or not? Is it worth the time to pursue the idea? Why or why not?

The critical reading checklist is a technique generally used by experienced design thinkers. However, beginners can also use the questions to challenge their team and get precise feedback.

TOOLS

- Paper
- Pens/markers

NOTES

DECLARATION OF CONSENT

Appropriate handling of information and data from conversations, observations and other research material is the basis of all work for every design thinker. Conditions are naturally made more difficult when data cannot be collected because organizational restrictions do not allow it.

Generally, we have our interview partners sign declaration of consent forms during projects. With participating individuals, these consent forms generate the necessary attention to make it clear that data is being collected. They also achieve transparency and trust, as the use of data is clearly laid out.

At a minimum, a typical declaration of consent includes the following information:

- First and last name of the individual from whom data is being collected;
- Purpose of the design thinking project;
- Statement allowing voice recordings or video data to be collected;
- Statement clarifying whether the data collected will be used anonymously within the scope of the design thinking project, or acknowledge the identity of the participant;
- In some cases, explicit permission to use pictures or video with the person's face will be used within the scope of the design thinking project;
- In some cases, explicit permission to use a person's correct name;
- Description of what will be done with the data collected at the end of the project; will it be saved (anonymously or with individuals' names) or erased?
- Contact details; telephone number, address, and/or email address;
- Signature with date and place.

Source

International Chamber of Commerce and ESOMAR, "ICC/Esomar International Code on Market and Social Research", 2008

TIPS AND TRICKS

- Find out whether or not there is already such a declaration of consent for use within your organization.
- Check whether or not the declaration of consent is also available in the participants' language(s).
- Always carry a few printed versions of the consent form (in the appropriate language) with you.
- Before you begin collecting data, inform your interview partner about the purpose and content of the project and the fact that you intend to collect data.
- Give your interview partner enough time to read the form.
- Have your interview partner sign the form. If possible, send him or her a copy after you return to your place of work.

3

THE DESIGN THINKING LAB

Creativity needs space! The interplay between organizational conditions and architecture leads to an environment that breeds creativity and innovation.

This chapter was written with the help of Michele Gaegauf.

CREATIVE SPACES

How does creativity arise?
What inspires us?
Where do new ideas come from?
What promotes or increases creativity and inspiration?

Good ideas cannot be forced; in general they also do not arise while one sits at a desk, thinking about a problem. They typically emerge from the subconscious and are therefore difficult to elicit. Inspiration does not come from oneself; instead external factors provoke it. Therefore, the right environment is absolutely able to foster creative work.

CREATIVE SPACES

Inspiration from the environment
Groves (2013) established that people, whether consciously or unconsciously, look for certain environments or establish situations in which they can best be creative and gain inspiration. Frequent examples are the following four places:

Nature: Fresh air, sunlight and the environment in nature fire up the imagination and bring about a sense of well-being and relaxation. The separation from one's familiar, often hectic environment lets the ideas flow.

Sports and physical activity: Active movement encourages blood circulation and spurs alertness. Repetitive, identical patterns of movement put the mind into a relaxed state. Interestingly, this relaxed state does not require sports; handwork such as knitting can achieve the same effect.

Bars and cafés: The visual stimulation and exchange with others brings about new thoughts and ideas.

Shower or bath: Like the famous example of Archimedes, many people report that the best ideas come to them in the bath or shower.

What all four places have in common is that they are spaces in which we can relax. In contrast to the desk at the workplace, here thoughts are not focused on the problem that needs to be solved. Rather, relaxation unconsciously leads us to a freer mental state.

In organizations, we can use these insights to foster the creative process by seeking to construct the aforementioned environments within a building. Google offers the best-known example of a creativity-fostering space within the workplace. Inside Google's offices in Tel Aviv, garden benches and tables rest in the shade of orange trees; the lighting is modeled after that in nature (inside!). In Zurich, there is a slide connecting two levels of a building, a small sports field, and a room styled after an old English library, with bookshelves, chandeliers, and plush chairs. Small meeting rooms are modeled after igloos and ski gondolas, embedded in a man-made winter landscape. In the water lounge, light effects and an aquarium simulate an underwater world. Couches and even a foam-filled bathtub invite us to free our minds and let the ideas flow.

Inspiration through communication
In addition to environment, communication is an essential element for inspiration and creative work. Often, the exchange of thoughts and plans with others brings about new ideas.

In general, we can differentiate between three forms of communication in organizations (Allen & Henn, 2011):

Communication for coordination: This form of communication is related to practical activities within organizations. It is determined by the structure, organization and culture of the company.

Communication for information: This type of communication serves the spread of information in organizations and is also determined by structure and organization.

Communication for inspiration: This comes from spontaneous encounters and interactions between employees, including with colleagues from other projects or departments. The conversations are relaxed ("hallway chatter") and interdisciplinary. Through this, such communication leads to new knowledge for both sides.

In organizations, emphasis is generally placed on communication for coordination and information. However, for a good communication flow, communication for inspiration is also important. For this to occur, the right conditions have to be established. These can be realized through the appropriate architecture (see Allen & Henn, 2011). The architecture needs to afford many opportunities for coincidental, unplanned communication — places where chance encounters happen, such as cafés and lounges. These should be set up so that multiple employees are able to (and want to) linger for sufficient periods of time. In addition, a space organized so that employees often leave their offices and frequently meet with a large number of colleagues fosters communication for inspiration.

At the Pixar headquarters, the entrance hall is large and open, and used as a central meeting point. The conference room, mailboxes and cafeteria are located off this space, which is called the atrium. Since even the restrooms are in the atrium, employees meet one another coincidentally and in passing on a daily basis.

The blueprint of the new planned headquarters of Apple in Cupertino shows a building laid out in a circle, with a park-like landscape in the center. The building plan is already focused around achieving as many inspiring encounters between employees as possible.

Creative spaces are places of chance encounters, of coincidental and inspiring exchanges, but also places of focus and stimulation.

They can be single areas, rooms, and also whole buildings. Creative spaces do not replace classic workspaces with desks and computers, but offer the organization and its employees important additional opportunities that a run-of-the-mill office cannot.

"Steve Jobs' vision was to design a building where people would interact naturally. He realized that when people interact and have fun, good things happen." (Capodagli, 2010)

"Steve Jobs' vision was to design a building where people would interact naturally. He realized that when people interact and have fun, good things happen."

(Capodagli, 2010)

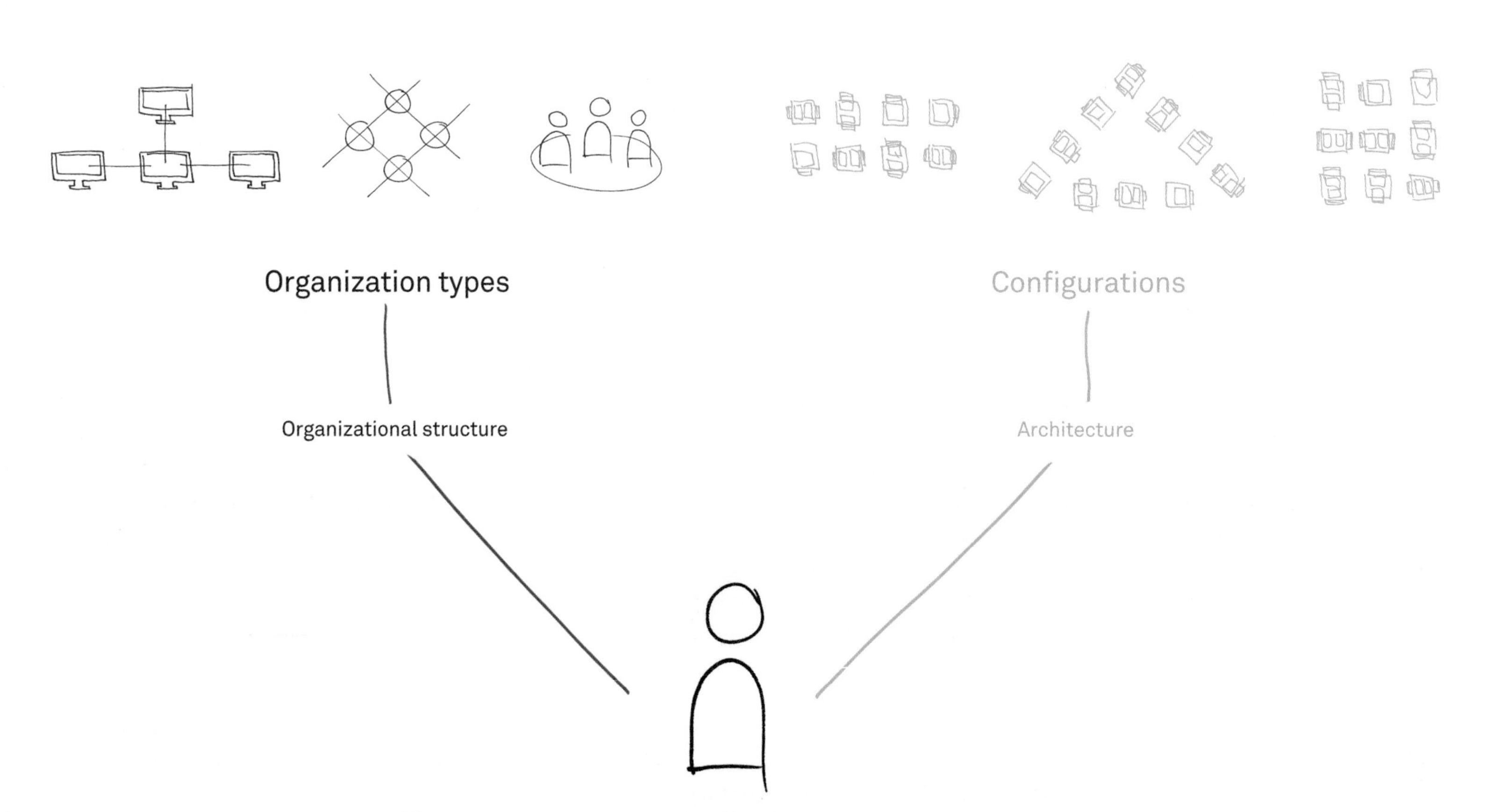

Possibilities for organizational configuration in line with Allen & Henn (2011)

GOLDEN RULES FOR DESIGN THINKING SPACES

When setting up design thinking spaces, one should first familiarize oneself with the general principles for creative work spaces. For this, there are a number of recommendations, the most important of which are outlined in "Make Space" (Doorley & Witthoft, 2012) and "What Makes a Great Workplace?" (Kahler Slater, 2010). From their own experience, the authors define rules for creative workspaces that are also relevant for design thinking, as follows:

1. Foster interaction in groups and teams
2. Stimulate creativity
3. Offer workspace choices
4. Focused and team workspaces must be available
5. Make co-creation spaces for work with third parties available
6. Offer large desks and lots of table space
7. Set up sources of inspiration such as libraries
8. Ensure natural and direct sources of light

In addition to these rules, our work on company projects and with teams at universities has resulted in the following "golden rules" for setting up design thinking spaces (Meel et al., 2010; Doorley & Witthoft, 2012):

Teamwork: The workspace for a design thinking team must inspire work in teams. Classical desks are not suitable for this. Instead, large, free surfaces on which the team can write and stick *Post-its* are required. Tables should be free-standing and accessible from all sides. Often, places for sitting are set up flexibly, with team members sitting on high stools as well as sofas. In many cases, it is not sufficient for design thinking teams to have only temporary work spaces. What is better, and often necessary, is long-term occupation and use of the space. Teams must be able to hang and display the results of their work on walls, windows, and other surfaces. A clean desk policy usually inhibits creative and team-oriented processes.

Workspace to focus: Even though design thinking is based largely on teamwork, team members need workspaces in which to be alone and undisturbed, so that they can concentrate on certain tasks. For example, patent research or a search for benchmarks requires time and concentration. This is usually difficult to find in team workspaces, which is why setting up focus workspaces should also be attended to.

Knowledge sharing: Areas for sharing and distributing knowledge, such as presentation surfaces and small stages, help teams to show their project status or results and thereby receive feedback. In their simplest form, these can be presentation surfaces equipped with a projector. Slightly more labor intensive to build are, for example, small stages with simple lighting.

Writable surfaces: If one visits a large German company, he or she often observes that the walls are decorated with artwork, but unfortunately there are no spaces for describing things in writing or in pictures. A single whiteboard is, however, too small for this. Teamwork, not just in design thinking, is based on the sharing of knowledge. This requires quick sketching of ideas, connections, and solutions. Writable spaces (everywhere!) are useful for directly retaining sudden inspiration, flashes of insight and solutions. To achieve this, it is even possible to make walls directly writable, and to use tables as whiteboards.

Roles and flexibility: "Everything has a purpose" is the saying in our design thinking workspaces. Design thinking teams need frequent workspace changes — whether for a short presentation that quickly requires a large area for roleplay, or the laying out and building of a prototype that needs additional space. Rooms and workspaces must be adapted to the needs of the team, not the other way around. To achieve this, furniture and other structures should ideally roll, rather than have fixed feet, to allow the team to rearrange the room to suit their own needs in a very short amount of time.

Prototyping space: Every design thinking team needs prototyping space — that is, space in which to build prototypes. The size is not immensely important, but materials for creation — such as paper wireframes, simple hardware prototypes and pads for scribbling — should be available and accessible. At the University of St. Gallen, we have a small corner with materials primarily intended for building service prototypes. Teams can help themselves to all the supplies they need, and after a few seconds begin building a prototype.

Creative space	Design thinking space
• Encouragement of interaction	• Furniture made flexible by rollers
• Stimulation of creativity	• Surfaces of tables and walls are writable
• Free choice of workspaces	• Large areas for collaboration
• Work areas for co-creation	• Focused work areas for undisturbed concentration
• Large tables and work surfaces	• Seven-second rule: tools and prototyping materials should be accessible within seven seconds
• Libraries as sources of inspiration	• Work areas are available to teams long term
• Natural and direct light	

WORKSPACES FOR DESIGN THINKING

Design thinking spaces are not just rooms dedicated to creativity. Spaces and rooms must be set up in a way that fosters creativity and enables inspiration. This is achieved by creating a workspace divided into various areas with different uses. This division can follow the flow of creative work, or pragmatically follow the usage of an area as one or more people employ it.

Organization in accordance with creative work exists, for example, in accordance with Groves, who distinguishes between areas to stimulate, reflect, collaborate and play (Groves, 2013). This classification correlates with the arrangement of workspaces necessary for different creative tasks.

We differentiate the opportunities for arranging workspaces in a similar mode, according to two dimensions (see Steelcase, 2014, figure on p. 232). The two workspace dimensions are individual or team and assigned or open.

Individual or team
Single workspaces are areas, such as offices and assigned workspaces, in which lone individuals can perform their tasks. These contrast with workspaces that can be used simultaneously by many people. An example is an-open plan office.

Assigned or open workspaces
Assigned workspaces are available to a sole person or team for a defined period of time. Open workspaces are available for a person or team to reserve temporarily, but are generally available to everyone in a company or organization over the long term. An example might be a café or a workshop room.

Both dimensions should be considered when arranging design thinking workspaces.

Quadrant 1:
Assigned individual workspaces
Design thinking teams need places to retreat to for tasks that demand high levels of concentration. This includes, for example, transcribing interviews or analyzing insights that emerge during need finding. These tasks also require that materials such as *Post-its* and printouts can remain laid out in the workspace. A clean desk policy would disturb the creative chaos on the desk. In practice, there are three fundamental architectural models for realizing these retreat spaces (see Meel et al., 2010):

Closed offices
Private and individual offices are generally a luxury in organizations. However, they offer employees the best opportunity to concentrate and to leave work materials on their desks. Particularly in projects with sensitive content, this kind of workspace should be considered, even if the workstation costs would be significantly higher. In general, closed offices are used only sometimes, and offer visual as well as noise protection. The primary interactions within the team happen in team spaces.

Workstations
Workstations are an alternative to closed offices. Most are separated from one another using soundproofing walls, so that focused work is possible. In contrast to closed offices, data confidentiality is more difficult to ensure in these more open environments. However, workstations generally offer an optimal working environment, since room dividers and a project's other materials can be quickly adapted.

1

2

1 Retreat spaces at Google (image: Google)

2 Design Thinking Loft at the University of St. Gallen (2013)

Open office

Open office interfaces/surfaces can also be used as room dividers, as long as the work surfaces do not need to be cleared daily and work areas are definitively allocated. Open offices have the advantage of enabling easy communication with other employees within a company (fostering inspiration). However, the disadvantage is that open office formats do not always offer design thinkers sufficient privacy to work on specific tasks that require concentration.

Quadrant 2:
Your own team workspace

This is an area in which the team can work privately for a longer period of time. This means that numerous surfaces are available exclusively for a certain team to use without the interference of others. This kind of workspace is important, as design thinking teams require surfaces and spaces to display, hang and store their work materials and results over an extended period (Doorley & Witthoft, 2012). A clean desk policy would damage the team atmosphere, as it would result in the constant removal of, for example, *Post-its* stuck on the wall during the design thinking process. At the University of St. Gallen, we allocate student teams with fixed team workspaces for a period of 9–10 months at a time. At the end of the project, they have become colorful tapestries. For the teams, however, the knowledge surrounding them is a decisive dimension of their working process. In practice, there are a number of ways to generate these spaces:

Closed team room

A room is assigned to a single team for an extended period of time — generally for the entire project. With this setup, the design thinking team can leave their work materials and utensils lying around as the project requires. The interaction of team members is very strong at times. Because it is closed, this type of workspace is recommended for projects in which confidentiality is of particular concern.

Open team area

Another variation of team spaces are open team areas. In this arrangement, a defined area of a larger room is made available to the design thinking team. Here, as well, the team can use walls, tables, and so forth over an extended period of time. However, they share the room itself with other teams. Curtains help to minimize noise and achieve sufficient privacy, making the room more usable. However, complete refuge from disturbances, and therefore confidentiality of project content, are difficult to achieve and ensure.

Quadrant 3:
Open group area

This kind of team space, which is open to use by all, can be established in numerous ways to serve many purposes, from a classic workshop or coffee corner to prototyping or presentation spaces. This kind of shared space is used above all for knowledge sharing, but also as a source of inspiration. Even coffee or break areas enable peers and colleagues, who in some cases may not have met before, to meet and develop new inspiration or encounter opportunities to collaborate (Meel et al., 2010). These spaces are generally only used temporarily and must be cleaned up afterwards.

Prototyping rooms and areas

Prototyping rooms and areas generally include machines, computers, and other work materials. In order to achieve optimal utilization, organizations generally make these rooms available to a large group of people. The size of the rooms and areas often correlates with the extent of design thinking used in organizations.

1

2

1 Team room at Google (image: Google)
2 Open group area at Google (image: Google)

Prototyping rooms can emphasize various activities, however in practice, most offer a mix:

Mechanical prototyping: These prototyping rooms make machines such as CNC mills, 3D printers and molding tools available to design thinking teams. Teams can use these rooms when making their ideas take the shape of prototypes. Generally, specific knowledge is required to use these machines properly.

Technology prototyping: These rooms have, for example, setups for simulations of three-dimensional worlds or challenging virtual realities.

Service and experience prototyping: Service and experience prototyping rooms can be rooms where, for example, parts of a store branch or service points of a given company can be modeled. A team might have a bank branch in mind, with all devices and furniture in place and ready for use. These rooms are effective for quickly and easily realizing and testing service prototypes.

Workspaces

Workspaces in these quadrants are, for example, the classic workshop rooms in organizations, which can be booked and used for short periods of time. These rooms can be considered for carrying out short meetings or conversations with users, as well as completing short-term tasks. Because of their short-term availability, these rooms are less appropriate for continuously working on long-term projects. Therefore, they generally only serve to supplement reserved spaces.

Hot desks: Hot desks are, in terms of their setup, open office spaces — that is, they do not offer employees an assigned workspace. In this sense, every employee can look for a new workspace every day (at least in principle). Relatedly, the workspaces must always be cleaned up after use. For this, employees are generally offered lockers or rollable containers. In practice, more than 10 employees generally work in these spaces.

Meeting rooms and meeting areas: Small and large meeting rooms and areas are good options for carrying out temporary workshops, training sessions or presentations. Additionally, they offer facilities for meetings with users and confidential conversations. Design thinking teams can reserve and use the rooms for particular events.

Brainstorming rooms: Brainstorming rooms are good for ideation. They usually feature inspiring objects. The walls are writable so that good ideas can be quickly communicated and recorded.

Meeting areas

Meeting areas are spaces in organizations in which employees can come together to share knowledge with one another.

Presentation areas: Presentation areas are a good option for demonstrating project results. Often, these do not just feature good presentation facilities, but sound systems and equipment for making videos. In some companies, there are even small stages or raised areas.

1

2

3

1 Prototyping room at Radicant Labs in Palo Alto (2014)
2 Design thinking lab at Deutsche Bank in Eschborn (2014)
3 Meeting room at the Swisscom BrainGym in Berne (2014)

Work lounges: Work lounges are community areas for two to six people in which employees can sit and address work-related issues together in an inspiring environment. Sometimes, these spaces are connected to cafés or areas where food can be purchased.

Meeting points: Meeting points are places in organizations in which informal meetings can take place.

Rooms for relaxation and inspiration
Rooms for general public use also include rooms for relaxation and inspiration. In large organizations, a number of variations can already be observed. "Healthy body, healthy mind" is an increasingly established motto for rooms devoted to sports and movement. For design thinking teams, these rooms offer an ideal change-up from the everyday routine, offering opportunities for both relaxation and new inspiration.

Cafés and break areas: These rooms, in which employees can enjoy small meals, snacks and drinks, can also be used as workspaces for short, informal meetings.

Games rooms: In technology organizations, there are sometimes pool tables in given areas. Particularly in the evening, team-building events can take place here in a relaxed atmosphere.

Gyms: Gyms are areas in which various strength and endurance sports and activities can be carried out.

**Quadrant 4:
Flexible individual workspaces**
The fourth kind of design thinking space is workspaces for single employees which, while separated, are often open for public use. The most frequently used are so-called "touch-down workstations". These can be flexibly used by all employees of the organization and are good for short activities and tasks. These stations are handy for checking emails, quickly surfing the web, or setting up appointments (see Meel et al., 2010). Another modern approach can be seen at the company Steelcase, which uses the "quiet" concept. Quiet spaces are small islands of calm, which can be set up in a countless variety of ways (Steelcase, 2014).

In addition, there are some well-known technology companies — for example, Google — that make thinking spaces available to their employees in addition to traditional workspaces:

Relaxation room: It might sound strange to set up rooms for relaxation. However, Google has rooms designated for this purpose, and we have done the same for our students at the University of St. Gallen. These rooms allow individuals space to unwind, read a book, or just think in peace. Who has not experienced sitting in an office and having to constantly deal with disturbances? These rooms prevent exactly these interruptions and enable concentrated thinking.

Study booths: Study booths are small corners for retreat which enable focused work via their separation from other areas. In some cases, old caravans converted into study booths can be found within organizations. Study booths are also ideal for telephone calls and conferences.

Libraries: Libraries can also be excellent places for both focused and relaxed work. Being surrounded by books establishes the right atmosphere for inspiration, calm and concentration.

The spatial design possibilities shown here are not comprehensive, but show the diversity of options available when setting up rooms and areas designated for design thinking work. What is important to note is that in practice, countless design combinations exist. These extend not only to a single room, but distribute the logic across multiple rooms within a building. In the following chapter, four frequently used combinations are presented.

Headquarters of Goodpatch Ltd. in Tokyo (2014)

Touchdown workstation

Relaxation room

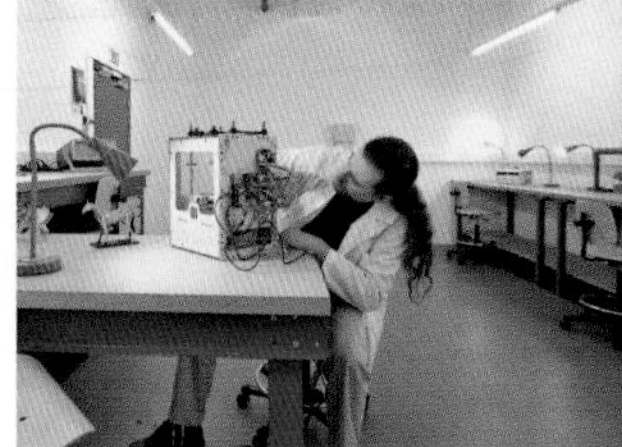
Prototyping area

Library

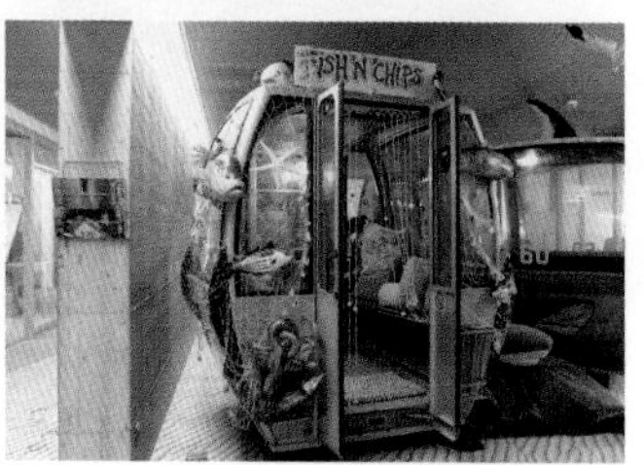
Study booth

Hot desk

Open

Flexible individual workspaces (Quadrant 4)

Open group area (Quadrant 3)

Single individual

Assigned individual workspaces (Quadrant 1)

Team's own work area (Quadrant 2)

Assigned

Closed offices

Workstations

Open office

Open team/project area

Large meeting area

Small meeting area

Large meeting room

Small meeting room

Brainstorming room

Presentation area

Work lounge

Meeting point

Break area

Café

Games room

Gym

Team

Closed team/project room

SETTING UP DESIGN THINKING WORKSPACES

When setting up design thinking spaces, achieving a good balance is often hard at the beginning. In many cases, design thinkers in companies must create new spaces while balancing existing infrastructure, available financial resources and the willingness of management to set up these new spaces. Through observations of numerous organizations over the past several years, four models for establishing design thinking spaces have been developed. The four models interact with and complement one another. In well set up organizations, all models can be found in parallel:

Ad-hoc design thinking spaces
Ad-hoc design thinking spaces use the existing infrastructure of the organization, such as whiteboards and flipcharts, to develop temporary spaces for design thinking. The existing material is complemented by easily procured and generally inexpensive materials such as foamboards.

Ad-hoc spaces are a good choice for organizations that want to try out design thinking on a small scale, or want to use design thinking only for selected projects. These spaces are also encountered in organizations that have integrated design thinking into their "architectural culture", and have put together ad-hoc rooms in parts of the organization not yet benefiting from high-level setups.

Design thinking lab
The design thinking lab is a room set up for long-term use in shorter-term design thinking projects, workshops and training. The lab typically features a modern furniture concept, and offers options for prototyping and storage of prototyping materials. Design thinking labs offer numerous companies a fixed and stable space in which design thinking can take place. This space is excellent when demonstrating to employees of an organization that design thinking is highly valued. However, because of space limitations (generally 100-200 square meters), such a room is not appropriate for a large number of projects being carried out simultaneously.

Design thinking floor
The design thinking floor extends the lab concept to the extent that it makes the space available to design thinkers substantially larger. The design thinking floor is like the lab, in that it is set up for long-term use, and long-running design thinking projects. Teams can reserve and use project rooms set up according to design thinking standards, and remain uninterrupted for long periods of time. Design thinking floors are of course associated with a higher investment by the organization. However, this step, which offers more available design thinking space, is necessary for the design thinking method to become an established approach in projects.

An example from Switzerland is the Swisscom BrainGym in Berne. Here, multiple floors of an old Swiss post office have been arranged according to the principles of human-centered design and design thinking. Employees and project teams at Swisscom can use these spaces for their projects on both a temporary and long-term basis.

Design thinking culture
For design thinking culture to become a central component of organizational culture, architectural requirements and specifications should also be adapted to it. Tech firms such as Google fully utilize the principles of design thinking when setting up new spaces. From floor to ceiling, architecture and infrastructure is built for the purpose of facilitating work in teams (see Walker, 2013).

From modern workrooms set up according to particular themes to relaxation rooms such as libraries and gardens, everything is possible when it comes to the design thinking spaces available to employees.

Within an organization, design thinking spaces of all shapes and sizes can be found. While in some cases an organization's headquarters may built according to design thinking culture and principles, other subsidiaries of the organization might feature a design thinking lab or floor plan. In practice, every organization can define its own degree of adaptation and realization.

Deutsche Bank design thinking lab in Eschborn (2014)

Ad-hoc design thinking

Ad-hoc team room

Design thinking lab

Small presentation area

Open team/project area

Storage area

Design thinking floor

Large meeting area

Closed offices

Large meeting room

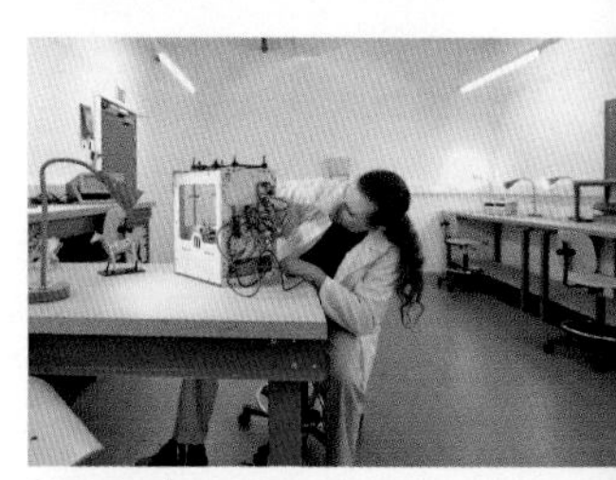

Prototyping areas

Library

Touchdown workstation

Level 1

Level 2

Level 3

Level 4

Brainstorming room

Café

Workstation

Design thinking culture

Presentation area

Study booth

Open office

Relaxation room

Small meeting area

Small meeting room

Work lounge

Games room

Gym

Break area

Meeting point

REALIZATION OF DESIGN THINKING WORKSPACES

AD-HOC DESIGN THINKING WORK AREAS

"Just do it!" is the motto for ad-hoc design thinking workspaces. Ad-hoc spaces are set up as temporary design thinking work areas. In practical terms, that means that classic workshops or workspaces are, in a short period of time, converted into design thinking workspaces.

Typically, rooms are de-structured and freed up so that the team can move freely in the space rather than sitting at a big desk. Here, it should be noted that even this adaptation of space can, at times, present a challenge. When working with a leading European chemical company, we were unable to remove large tables from rooms, since they were fixed to one another, and to the floor, with screws. Sometimes it is necessary to have facility management employees at the organization temporarily remove tables and chairs so that the team has the necessary freedom of movement.

Furthermore, it is advisable to use foamboards in ad-hoc design thinking rooms. These lightweight, highly transportable foam or cardboard walls do not cost much, but greatly increase surface areas for *Post-its*, drawing and writing in a very short period of time. One thing to consider is that *Post-its* do not stick to every wall surface. Unfortunately, they all too often fall off. The surface of foamboards solves this problem!

Materials necessary for prototyping can be easily stored in rollable containers in ad-hoc design thinking rooms. An organization in Switzerland used old airplane cabin crew service carts with rollers for this purpose. They do not need much space, are practical, and look good.

USAGE SCENARIOS

- Organizations that want to try out design thinking
- Organizations that want to use design thinking only for selected projects (small number of projects with a short running time)
- Organizations that want or need to carry out design thinking projects in areas with insufficient design thinking setup and materials

Ad-hoc design thinking room, Royal Bank of Scotland Technology Solutions Center, Edinburgh (2014)

REALIZATION OF DESIGN THINKING WORKSPACES

DESIGN THINKING LAB

The design thinking lab is a room within an organization set up for long-term use and made available to teams or groups. The room typically features the most important objects for the setting up and organization of a design thinking room and is up to 200 square meters in size. The room is arranged to serve a few short-term design thinking projects, while also providing a space for training and workshops.

Workshops and training
The most frequent use for design thinking labs are workshops and training. The rooms offer ideal working conditions for short format events, particularly in terms of infrastructure and availability of work materials. Teams and groups can book and use the lab for short periods of time. After the events, workspaces are returned to their original state. That means that the results achieved need to be documented and brought along when leaving the lab.

Short-term projects
Other uses for the lab include short-term projects. Because of the limited space available, labs are seldom used for long-term projects, as this would prevent them from being available to other teams. For short-term projects, part or all of the room is reserved for the project team. Results can be left out on a workspace for a longer period of time, and do not have to be constantly cleared.

Organizationally speaking, the design thinking lab in organizations is generally managed by a responsible party. This person or group organizes the use of the room, the borrowing and return of material, and ensures orderly use.

USAGE SCENARIOS

- Organizations that want to use design thinking on a temporary basis for selected projects, and wish to carry out training in a lab
- Organizations that want to support design thinking projects in external locations with a professional setup, without this location being forced to completely rework their architectural design concept

Design thinking lab at Deutsche Bank, Eschborn (2014)

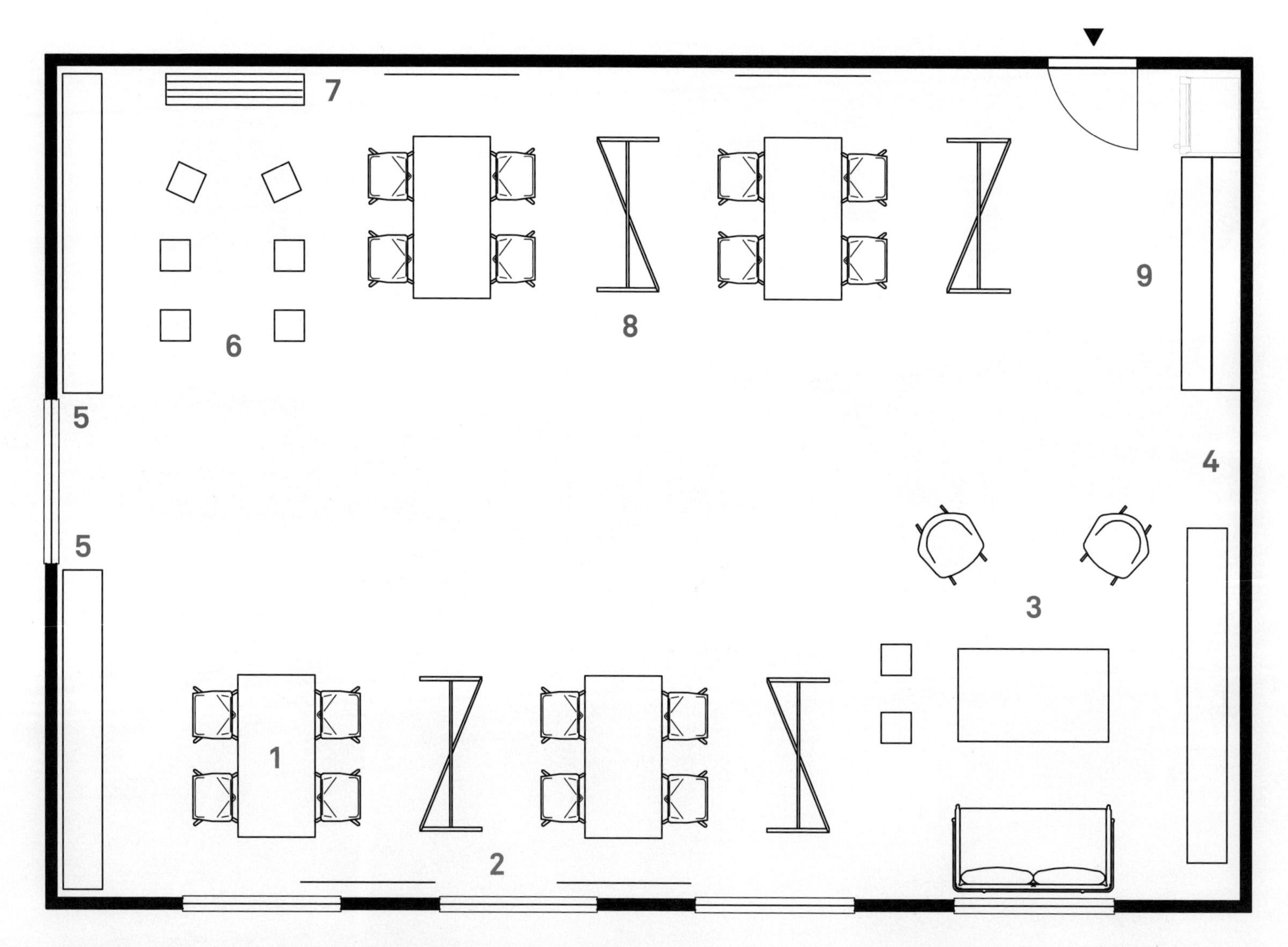
7
6
8
9
5
5
4
3
1
2

Architectural plan of a design thinking lab

1. Team room: Standing tables with strong surfaces and wheels have proven themselves to be an excellent resource for design thinking labs. There are even some models with surfaces that can be used as whiteboards. Barstools should also be available, and work well with the standing tables.

2. Foamboards: These light and mobile surfaces offer the team a place to put their ideas and foster group work. The more foamboards, the better. Because of their light weight, they can be leaned against walls, stools and chairs, and other objects.

3. Library: A couch, a variety of chairs, and possibly an easy-chair should be arranged into a comfortable sitting corner, where design thinkers can hang out and share ideas over coffee.

4. Free storage space — library: A shelf with lots of surfaces and space, as well as open access, serves as a small library. Inspiring magazines and books can be stored here, as well as a few curious and/or fascinating objects.

5. Free storage space — prototyping: Tools and materials for building prototypes can be put away and stored here.

6. Presentation area: This area serves as a presentation platform. A projector, a screen or surface for projecting, a lectern and chairs should also be available.

7. Additional/alternative storage space: Walls can be used as storage spaces, for example for additional foamboards.

8. Whiteboards: Like foamboards, whiteboards serve as work surfaces and room dividers simultaneously. Some models also have wheels and can be freely placed anywhere in a room.

9. Café area: A small cooking and washing-up area should also be available, depending on the space and the nature of the setup. We recommend a coffee machine, a kettle, a microwave or stove and a refrigerator.

REALIZATION OF DESIGN THINKING WORK AREAS

DESIGN THINKING FLOOR

Organizations that view design thinking as part of their work culture often set up a design thinking floor — that is, the conversion of an entire floor of a building for working in line with design thinking principles on a permanent basis. In contrast to a design thinking lab, the additional space offered by the design thinking floor enables multiple teams to work simultaneously on projects in the same rooms over a long period of time.

Design thinking floors offer, in addition to the infrastructure of the ad-hoc rooms and the design thinking lab, areas specifically designated for prototyping and breaks. They also offer more extensive options and spaces for retreat and relaxation, as well as the presentation of results.

The exemplary room plan for a design thinking floor mapped on p. 242 shows a basic layout within a floor area of 300 square meters. In contrast to a design thinking lab, here it is possible to better divide workspaces from one another with the help of movable walls. This helps ensure confidentiality, while reducing interruptions and minimizing distracting noises from other teams. In addition, the space features a sitting area and library. It offers opportunities for retreat, where one can read books or articles in peace, but also do focused work on a laptop. The design thinking floor is completed by a large presentation area in the middle, a meeting room and four closed offices with individual workspaces for focused work.

USAGE SCENARIOS

- Organizations that want to anchor design thinking as a central part of their working culture
- Organizations that want to make design thinking rooms available for a large number of projects

The BrainGym at Swisscom. The BrainGym offers Swisscom employees ideal conditions for carrying out human-centered design methods (2014).

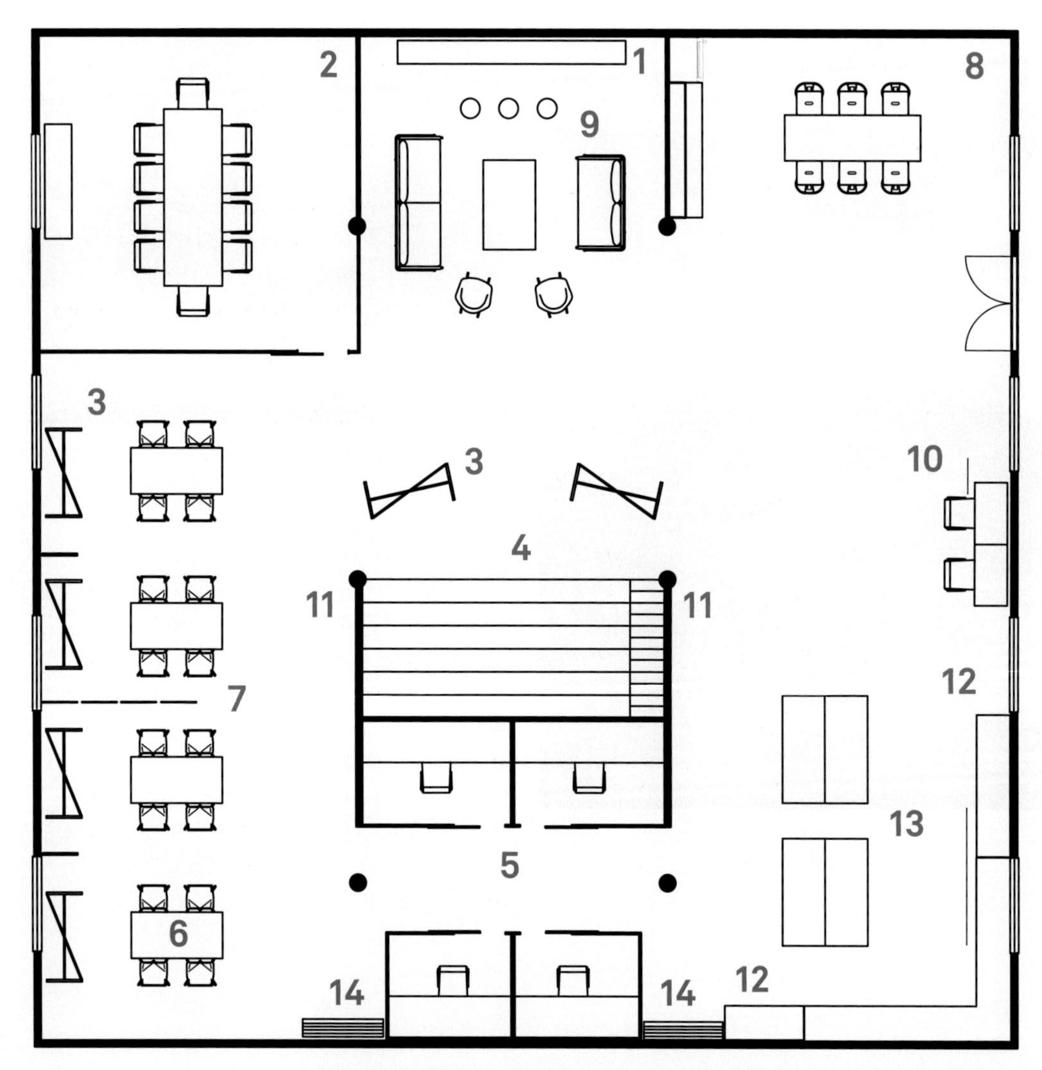

Architectural plan of a design thinking floor

1. Open storage space
An open-access shelf with a lot of surface area serves as a small library. Inspiring magazines and books and selected interesting and inspiring objects can be stored here.

2. Large meeting room
This room can be used for meetings.

3. Whiteboards
In addition to foamboards, these simultaneously serve as workspaces and room dividers. Some versions have rollers and can be freely placed around the space.

4. Presentation area
Here, steps are often used to make a stage for presentations. Similarly, chairs and other objects can be used for presenting ideas and prototypes.

5. Private offices
Individual workspaces should be made available in a closed room so that employees can carry out focused work in a quiet environment. These rooms are occupied according to need.

6. Team room
Standing tables with a robust surface and rollers have proven their utility in design thinking lofts. Some versions also have surfaces that can be used as whiteboards. Barstools go well with these tables, and should also be made available.

7. Flexible dividing walls
Walls, curtains, or folding screens can also serve as room dividers, offering teams the option to separate their workspace. Ideally, these would be chosen and arranged so that they could also be used for working upon.

8. Café area
A small area for cooking and washing up should also be available so that employees and team members can prepare food. Depending on space and resources, we recommend a coffee machine, a kettle, a microwave or stove, and a refrigerator.

9. Library
A couch, various chairs, and perhaps an easy-chair should be comfortably arranged in a corner, for passing the time and exchanging ideas over coffee.

10. Touchdown workstations
Printing, checking emails, research... Some computers should be available for these kinds of activities.

11. Brainstorming area/surfaces
Walls with whiteboards or writable sections are enormously useful in design thinking environments. They serve as a basis for brainstorming and synthesis.

12. Open storage space
Tools and materials for building prototypes can be stored here.

13. Prototyping room
Work benches or tables for prototyping should have large, robust surfaces. Rollers offer helpful flexibility, and have proven a very useful feature in standing tables.

14. Foamboards
These light and mobile surfaces offer the team space for ideas and group work — the more foamboards, the better. Because of their light weight, they can be leaned against walls, chairs and other objects.

REALIZATION OF DESIGN THINKING WORKSPACES

DESIGN THINKING ORGANIZATION

Anchoring design thinking as a central part of organizational and working culture requires dedicated alteration of workspaces. As part of this development phase, organizations actively change all their rooms and spaces in accordance with the fundamental working principles of design thinking.

Study booths, team workrooms, work lounges and hot desks are available in large numbers. According to our experience, in addition to workrooms, organizations also make rooms for retreat and balance available to their employees. These may consist of small fitness centers that can be easily and quickly used over the course of a workday, but there may also be spaces in which to take breaks. Through this, design thinking organizations achieve an environment in which employees may not only work, but also live. Some tech companies provide their employees with everything they may need — from breakfast to dinner — and thereby create a free space in which their employees can completely focus upon their work (Walker, 2013).

USAGE SCENARIOS

Organizations that consider design thinking a central cultural component

1

2

3

1 Work lounge at Google (image: Google)
2 Creative space at Goodpatch, Tokyo, Japan
3 Meeting room in the form of a library, Google, Zurich (2015)

DESIGN THINKING MATERIAL LIST

The material list is a compilation of what we view as the most important materials for use in all kinds of design thinking rooms, spaces and buildings. Here, we have compared the lists of various labs and also incorporated our own experience. Of course, this list can be extended and elaborated according to requirements. After the composition or construction of a design thinking space, the question of what should be acquired is often the first to arise.

1. *Post-its:* Design thinking can no longer be imagined without *Post-its*. They are integral to the simplification of teamwork and structuring complex material. In brainstorming, *Post-its* help with the quick expression, retention, and communication of ideas with team members. When transcribing interviews, they help the team retain essential information in a striking way. We recommend keeping different sizes on hand.

2. *Foamboards and pins:* Foamboards are excellent for putting up and making order out of *Post-its*, and also other notes and even small objects. Every team should have two to three foamboards. Projects with longer running times — over three months — can quickly require up to 10 boards.

3. *Markers:* Writing utensils are clearly critical. Sharpies and the Edding Permanent Marker 1300 have differentiated themselves as particularly good for *Post-it* writing. The markers have a thick enough tip that writing can be read from far away, but they are fine enough that enough text can be written on a *Post-it*. These, and whiteboard markers, should be available in sufficient numbers.

4. *Index cards:* Cards to write on and hang up on walls and other surfaces are obligatory. Index cards are also good for prototyping!

5. *Paper:* Sheets of paper in different colors and sizes, as well as packing paper for large sketches and prototyping, should also be available.

6. *Adhesive:* Glue sticks, adhesive film, tape, hot-glue guns, duct tape and adhesive labels and dots should be standard.

7. *Prototyping materials:* A foundational stock of materials for prototyping should always be available. This includes pipe cleaners, *Lego* blocks, modeling clay, styrofoam, aluminum foil, wood, cardboard in different forms and colors, as well as construction paper. Tools such as scissors, carpet cutters, saws, mat for cutting, rulers, measuring tape, cordless drills, pliers, nails, screws, staples, clips and sandpaper also belong in a well-filled prototyping box. Clothing for costumes should also be available for roleplay.

POST-ITS

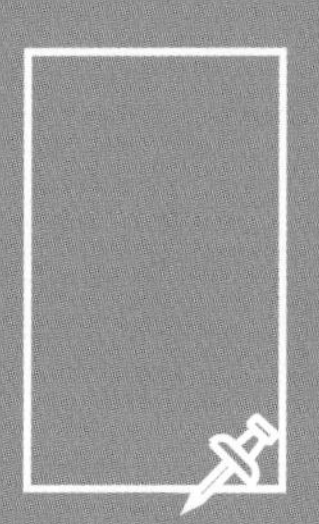

- *Post-its* in different colors and sizes
- *Post-it Super Sticky Meeting Flip Chart*

FOAMBOARDS AND PINS

- Foamboards
- Clear bulletin board pins

MARKERS

- Sharpie Fine Point (various colors)
- Edding 1300 permanent markers

PROTOTYPES

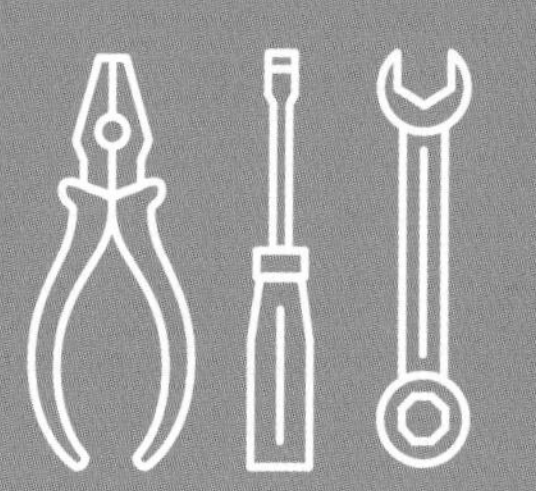

- *Lego* blocks, modeling clay, styrofoam
- Pipe cleaners, aluminum foil
- Wood, cardboard, boxes
- Materials for costumes
- Carpet cutters, cutters, scissors, mats for cutting
- Rulers, measuring tape, cordless drill
- Saw, hammer, screwdriver
- Staples, clips, pliers, nails, screws, sandpaper

INDEX CARDS

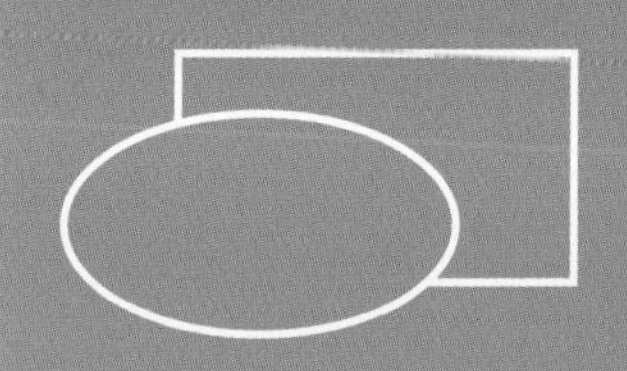

- Various sizes and colors

PAPER

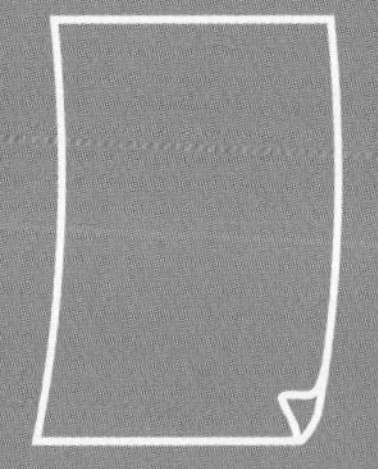

- Packing paper
- Colored drawing paper/ cardboard
- Selection of newspapers and magazines

ADHESIVE, TAPE, LABELS

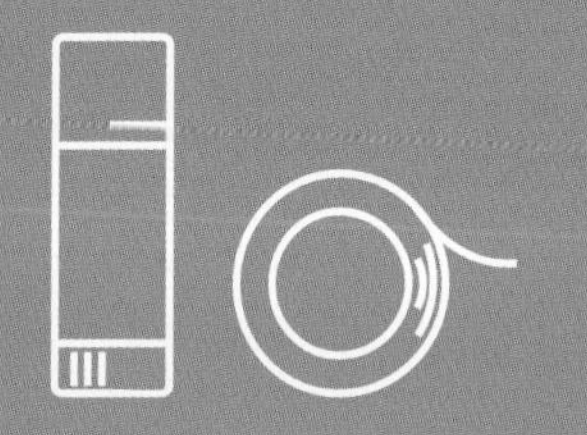

- Glue sticks, hot glue
- Masking tape, tape
- Colored adhesive dots (19 mm)

DESIGN THINKING TOOLS

Here, we note the most important groups of tools that should be in a design thinking space. In discussions in organizations, one might point out, for example, that stopwatches are not necessary and that a cellphone would work just as well. Do not let yourself be confused or swayed — these materials are necessary and should be procured! Being able to see the physical clock helps the team work in a timely and synchronized way during the different work phases.

1. *Stopwatch:* Timekeeping is important for numerous group interactions so the team does not lose track of time planning. Most design thinking teams in Europe use the *Time Timer* from www.timetimer.com, which enables the team to see the remaining time for a brainstorming session or PechaKucha presentation. The large model is ideal, as all team members can read it simultaneously. The small model is the perfect accompaniment when traveling.

2. *Flipcharts and foamboards:* These can be used for presentations and are absolutely necessary for work in teams, as they can freely be covered with *Post-its* according to the team's needs. Flipcharts are useful for larger sketches, and foamboards are good for putting up cards and even small objects. Do not forget the appropriate markers and pins!

3. *Recording devices:* Use digital photo and video cameras to record activities, research and test documentation and, in some cases, to bring these to testing participants. These do not need to have all the features of a professional setup. Small devices are sometimes better, as they are more user-friendly. Good still cameras, which remain practical despite their numerous other functions (including the ability to record video), are the Canon *PowerShot* or Nikon *CoolPix* series.

4. *Computer with loudspeakers and a printer:* These are critical for internet research, saving and playing recordings, presentations, and for generating and saving important documents such as declarations of consent. It is important to consider that video sequences can be easily edited on computers. Apple has proven itself to be the gold standard in this regard, since the video editing software *iMovie* is used ubiquitously.

5. *Devices for presentations:* A projector and projection screen with a (wireless) computer connection should also be part of the design thinking layout.

6. *Trash can, vacuum, broom, dustpan:* There must be order. Even though work materials can, and even should, lie around over the course of the project, the area should nevertheless be regularly tidied up to ensure there is room to work. Multiple large trash cans are an absolute necessity!

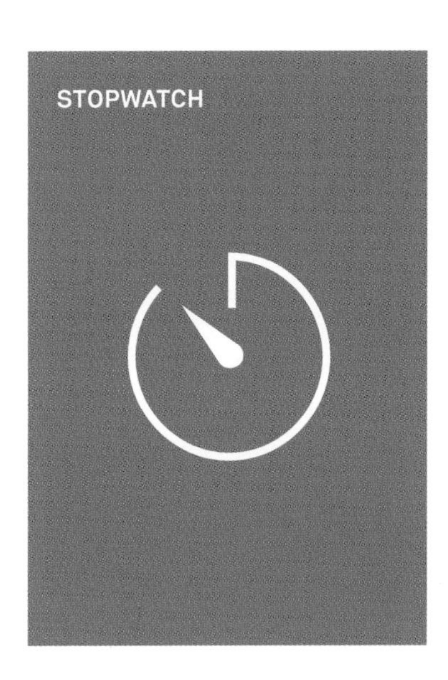
STOPWATCH

GARBAGE CAN

RECORDING
• Camera for need finding, presentations and prototyping
• Audio recorder for need finding

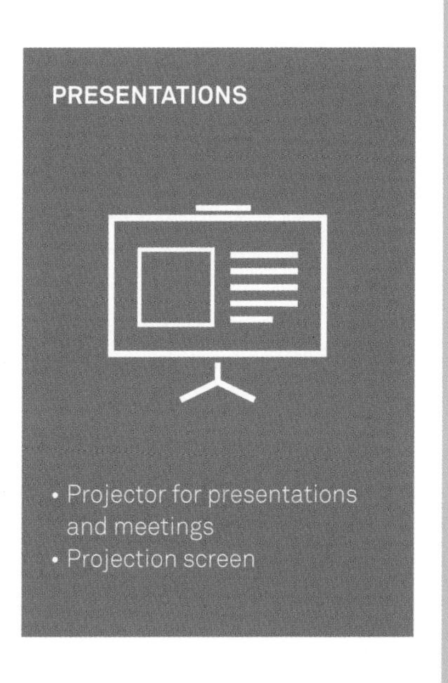
PRESENTATIONS
• Projector for presentations and meetings
• Projection screen

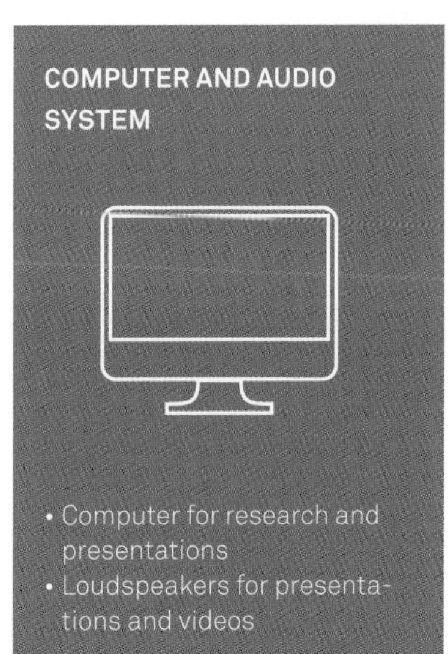
COMPUTER AND AUDIO SYSTEM
• Computer for research and presentations
• Loudspeakers for presentations and videos

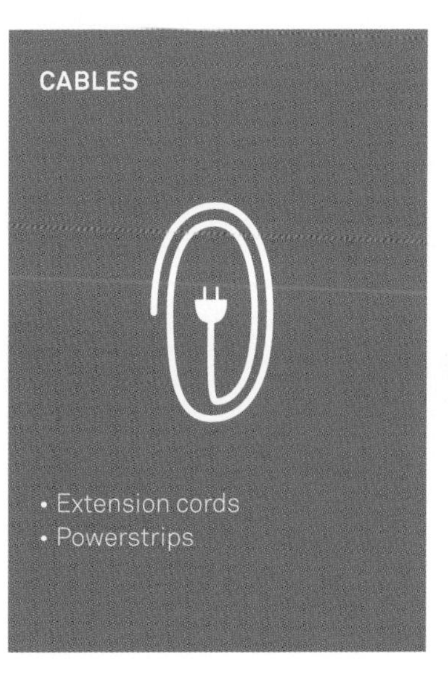
CABLES
• Extension cords
• Powerstrips

CLEANUP
• Vacuum cleaner, broom, brush and shovel, for cleaning up space and materials after use

FLIPCHART EASEL
• For meetings, presentations, brainstorming activities and visualizing

DESIGN THINKING FURNITURE

Most design thinking spaces look like little apartments, and this is good! The goal is not only to create a productive workspace, but also to ensure design thinking team members feel at home. As previously described, they should include sofas, sitting corners and green plants.

1. *Tables:* For teamwork and prototyping, there should be multiple tables available. The height should be adjustable, or the tables should be high enough that team members can work standing up. The tables should have robust surfaces that can be wiped down, and should ideally be writable. They should be movable, with rollers, in accordance with the team's needs. To this end, the company System 180 has devised some clever solutions.

2. *Whiteboard:* For presentations and teamwork, whiteboards are great. When procuring them, make sure they easily moved and stacked against one another to save as much space as possible. This is where cheaper and more expensive whiteboards differ most from one another.

3. *Chairs or barstools:* The height of the choices for sitting should be determined according to the tables. Here, the guidelines for whiteboards also apply: Chairs and stools should be easily stackable. It is also important to consider whether everyone can tolerate and use the stools. Unfortunately, this is not a given.

4. *Shelves:* Shelves constructed from wood or metal are best for storing materials. Make sure the shelves are open, so it is easy to identify their contents. Otherwise, notes on shelf doors are useful for quickly determining the contents of a given drawer or cupboard.

5. *Storage boxes:* Boxes made of metal, wood, plastic or wire are good for storing diverse materials and can be easily stashed on shelves.

6. *Carts:* These ensure that materials and tools are always ready for use — everywhere in the design thinking area. Recently, discarded service carts from airlines Swiss and Lufthansa have become available for purchase. These work fantastically and look great.

7. *Sofa:* Ideally, also equip this with rollers.

8. *Furniture rollers:* If you put furniture on rollers, the work area can be quickly rearranged in accordance with requirements.

9. *Environment:* In addition to furniture, plants, humidifiers and natural light are important to create a positive work environment.

TABLES

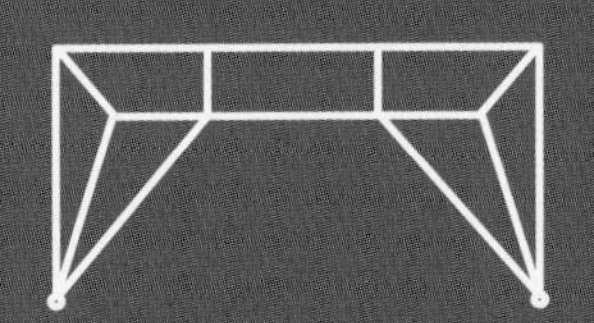

- Standing tables with rollers and adjustable height
- Tables should be robust and (if possible) have writable surfaces

WHITEBOARDS

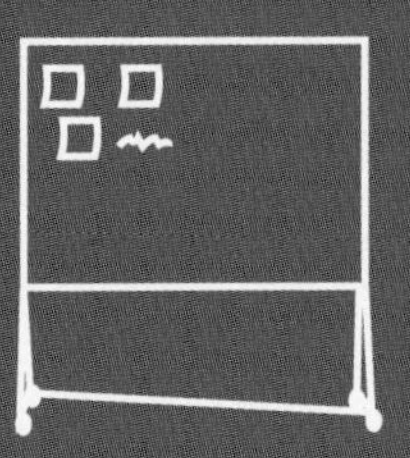

- Whiteboards for presentations, brainstorming, workshops or teamwork

CHAIRS OR BARSTOOLS

- Stools in different sizes and heights (for normal and standing tables)
- Stools and chairs should be stackable

SHELVES

- Shelves for storing work and prototyping materials

STORAGE BOXES

- Boxes for practical storage of materials and other tools and utensils for work

CARTS

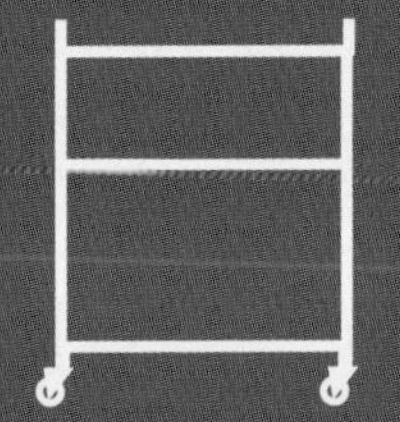

- Carts for quick and easy movement of work materials

SOFA

- For the library, or places in which meetings and conversations are carried out
- Also with rollers, so that it can be freely moved and placed within the room

INTERIOR DESIGN

- Green plants
- Natural sources of light and indirect light
- Humidifier

4

APPLICATION IN ORGANIZATIONS

The implementation of design thinking progresses via pathways of transformation.

DESIGN THINKING IN EVERYDAY PROJECTS

Over the past decade in Europe, a number of organizations have worked to implement design thinking in their organizational cultures. At the forefront is German company SAP AG. In Switzerland, Swisscom began its first collaborative projects with the University of St. Gallen and Stanford University as far back as 2005. As authors, we have been able to take part in over 60 projects during this time period, as both researchers and consultants.

"One can say so many positive things about design thinking — its operating principles and their impact. But to really understand it, one has to have experienced it firsthand."

(Marcel Plaum, Fraport)

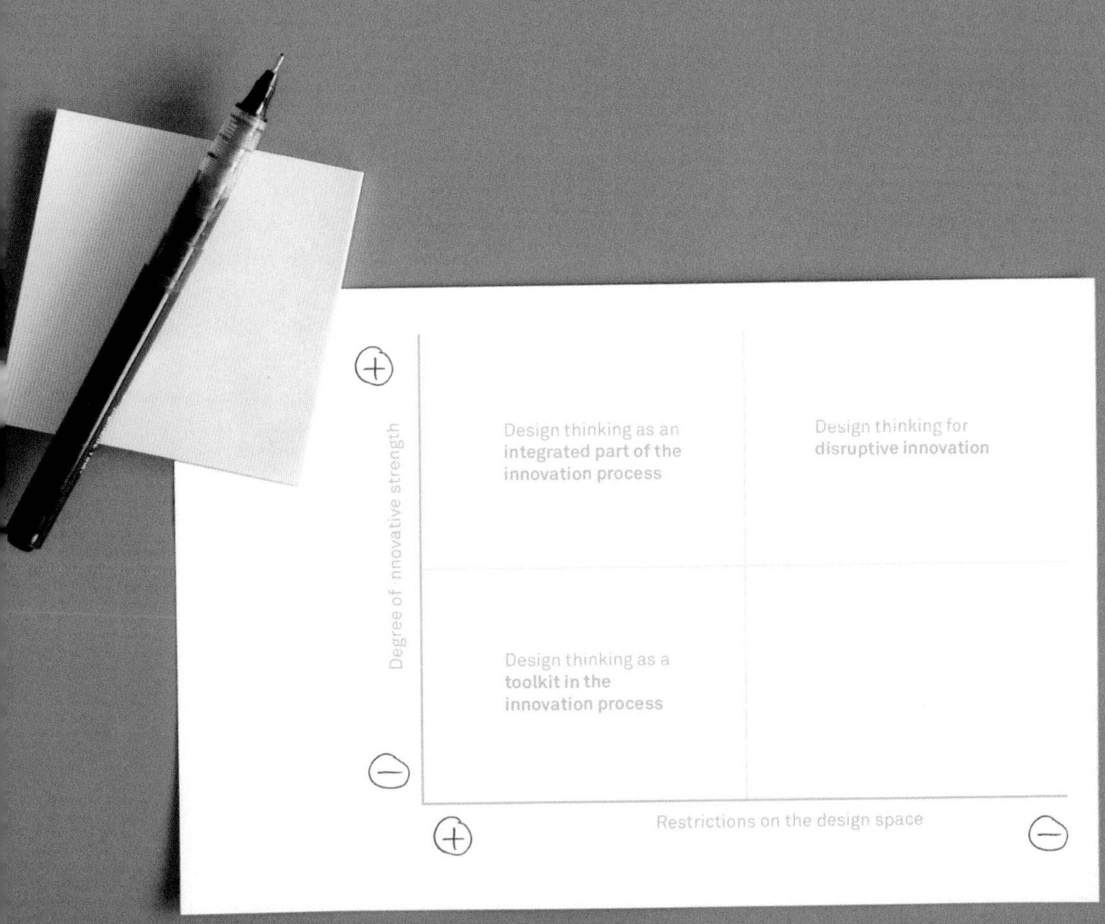
Design thinking as an integrated part of the innovation process
Design thinking for disruptive innovation
Design thinking as a toolkit in the innovation process
Degree of innovative strength
Restrictions on the design space

From our experience, two dimensions related to the implementation and operational everyday usage of design thinking have crystallized as being important (see matrix on p. 259):

Degree of innovation: Depending upon the project's question statement and how this fits into the organizational context, highly innovative and creative solutions are required. Nevertheless, there still exist many question statements that do not demand a high degree of innovation. These question statements focus "only" on improving, and further developing existing products. Design thinking can be used in both contexts, but it requires adjustments to be useful.

Restrictions on the design space: The second dimension that impacts the use of design thinking and its configuration are restrictions on the design space. If project-specific factors restrict the use of design thinking, then its usefulness as an innovation method becomes limited. In such cases, its usage should be adapted. If the number and significance of restrictions are low, the project team has more freedoms in the search for, and implementation of, new solutions. This also leaves room for disruptive innovations.

From the innovation matrix on p. 259, we can begin to see three operational variations of design thinking. They are detailed below:

Design thinking as a toolkit in the innovation process: Design thinking is selectively used only within parts of projects. For example, a development project with clear parameters and a low degree of uncertainty from user interviews (see need finding pp. 27 – 29, pp. 96 – 99) and tests would profit from design thinking. Similarly, in such projects, low-resolution prototypes within the team quickly reveal the fragilities of the final design. For more detail, see pp. 32, 148, 178.

Design thinking as part of the innovation process: In this form, design thinking is employed as guidance for divergent phases (see p. 36) — that is, up to the funky prototype (see pp. 36, 46). The later phases are typically executed via other methods, such as SCRUM (www.scrum.org). In this mode, the application of design thinking is limited to a few days, and perhaps mapped out over the course of short projects or workshops. For more detail, see p. 60, 69, 262.

Design thinking for innovation and customer orientation: On the basis of low restrictions on the design space and high requirements for the strength of innovation, design thinking is implemented in its complete form, as described in this book.

Furthermore, we have found that it is at times useful to rename phases of the macro process (see pp. 36 – 49). Why? In purely pragmatic terms, names like "dark horse" or "funky" prototype are not self-explanatory. It is not clear to everyone what processes lie behind these names. Therefore, together with project practitioners we have come up with the following alternative terms:

Dark horse prototype: Visionary prototype
Funky prototype: Integrated prototype
Final prototype: Proof of concept

The quintessence of these single phases remains intact, as do the deliverable results. However, the advantage is that, contrary to the term "dark horse", more practitioners in organizations can understand and envision a visionary prototype. The same is true for integrated and proof of concept prototypes.

DESIGN THINKING IN EVERYDAY PROJECTS

DESIGN THINKING AS A TOOLKIT

Viewing design thinking as a big toolkit is an approach that organizations take to more centrally integrate its cultural components into their existing organizational processes and routines. While "pure" design thinking projects concentrate above all on new products, services and business models, most projects in organizations are focused upon incremental improvements and change. These can also be enhanced using components of the design thinking method.

For example, while building software products, it can be worthwhile to complement classical processes of so-called requirement engineering with need-finding methods (Vetterli, Brenner, Uebernickel & Petrie, 2013).

Product development in organizations can also benefit from prototyping methods. We have been able to use low-resolution prototypes from the design thinking toolkit to achieve good results with financial service providers. Complex consulting situations and concepts can often be rapidly perceived with the help of prototypes. Building, but also getting the opportunity to connect a concept to a concrete object quickly and efficiently, leads to further discussion. Strengths and weaknesses can be more clearly identified and discussed together with users during testing.

To use design thinking as a toolkit, however, requires expert knowledge on the tools of the organization. Employees have to be trained and educated. In practice, training days in which one or two methods are theoretically and practically explained and then practiced can work well.

The advantage of this approach and understanding of design thinking is that small and visible results can be achieved. Because the usage area is limited, the complexity is generally low. The risks and costs for the organization are typically straightforward and manageable.

A disadvantage is that the weak use of design thinking clearly does not have the same impact as the complete employment of the methods. Leaving out the divergent phases generally means that an innovative jump may not happen. In this context, it needs to be clearly communicated to employees that they will be using only design thinking tools, not design thinking methods.

DESIGN THINKING AS PART OF THE INNOVATION PROCESS

Another form of implementation we have often observed is the integration of design thinking into an organization's innovation process. Instead of using only some design thinking tools, parts of the design thinking process are intermingled with components of the existing innovation process.

Numerous organizations set the goal of above all profiting from the divergent phases of the design thinking process. A frequently found implementation variation is the upstream integration of design thinking with agile development methods such as SCRUM. In this form of use, the first phases of the macro-process are applied, typically through to the integrated (funky) prototype. The process is then concluded via an agile approach to development.

Using this procedure, many organizations hope to help their project team members think outside the box during the early phases. In converging phases of the project, they hope to develop new solutions despite restrictions.

The potential forms of this implementation family are relatively diverse, and guided by organizational needs:

Workshops: For the divergent phases of projects, multi-day design thinking workshops can be carried out with team members. The advantage is the enabling of compact workshops with the chance to integrate substantial team dynamics. Nevertheless, it is important to ensure that expectations for workshop results are not too high. Particularly when there is insufficient time for solid preparation, the richness of results is often sharply limited. In many cases, a few days or hours are not sufficient for solving complex problems and projects. In these cases, carrying out mini-projects is recommended.

Mini-projects: “Mini design thinking projects” focus upon the early phases of need finding, synthesis and ideation, and provide inspiration and ideas for the agile methods used in later stages. These projects generally have a short running time of a few weeks (dependent upon the number of interviews and observation). In comparison to workshops, the advantage is that richer, deeper results are made possible.

DESIGN THINKING IN EVERYDAY PROJECTS

DESIGN THINKING FOR USER ORIENTATION AND INNOVATION

In its comprehensive form, design thinking is used primarily when organizations have substantial freedom in relation to both the degree of innovation and design space. Generally, these projects focus on (new) developments for products, services and business models.

For example, today some sectors (such as energy) are faced with the question of how their products and services will look in the future, and which business models will add value for users, shareholders and employees. In these situations, the comprehensive use of the design thinking process is the best way to tackle these challenges.

In such cases, organizations typically designate specific project teams which use the design thinking method to tackle these question statements over a time frame of several weeks or months. Characteristically, both the divergent and convergent phases of the design thinking macro-process are carried out to their completion. This ensures the integration of the customer and user up until the final prototype.

The result is usually a prototype of a product, service or new business model that has still yet to run through a classic development process. The advantage of this approach is that the complete use of design thinking methods achieves a comprehensive integration of the user, and the entirety of design thinking principles ensures impact.

LEVELS OF ORGANIZATIONAL TRANSFORMATION

Parallel to implementing design thinking in the everyday operation of the project, levels of design thinking implementation at organizational and strategic layers have also been built up. An organization's goal must not always be to achieve the highest possible degree of design thinking adaptation. Depending upon the individual market conditions and the embedded strategic orientation, smaller or preliminary objectives can also support the organizational objectives.

Here, we briefly describe four patterns we have identified over the years:

Level 1 — Awareness
This first level represents the entry of design thinking into the transformation of an organization. The objective is to achieve an understanding of customer- and user-orientation, creativity, and an iterative work process among employees. The concrete measures tend to be:

Town hall meetings: Town hall meetings are events for employees that are typically large format (100–200 participants). External guest speakers are invited and typically speak on projects within other organizations, new innovative methods, or innovation itself. With the help of the town hall meetings, a general feeling for innovation and, above all, how innovation is achieved, can be generated among employees.

Individual employee training: At this level, individual employees are trained as promoters of the design thinking method. Training is often held externally, at universities or other facilities. Less frequently, employees are trained and participate in organizational pilot projects in parallel. This training variation is typically more sound and intensive, however it requires more time and human resources.

Situative projects: Design thinking projects are carried out in situative manner. Generally, because of an internal deficit of design thinking knowledge and resources among employees, they are led by external consultants or other capacity builders.

Spaces: Projects at this level are generally carried out in ad-hoc spaces (see p. 234).

Level 2 — Experimental
On the second level of development, design thinking is already a method perceived to have the potential to become anchored within the organization, and is classified accordingly. Design thinking is understood as a method that offers potential for solving selected problems, projects and tasks. In the foreground is the wider and at times prototypical application of design thinking:

Training: Beside the promoters, who are trained at level 1, selected project collaborators and directors are trained in design thinking methods. At this level, the focus is building a strong connection between design thinking and existing methods in the organization, and demonstrating the value of design thinking.

Projects: The use of design thinking as a toolkit or method in the innovation process is more pronounced than at level 1. Previously trained employees represent an increasing majority of those working on the project, so that methods are experienced and propelled forward by internal employees.

Awareness	Experimental	Catch-up	Innovation
• Town hall meetings • Training of single employees • Situative projects • Spaces	• Local innovation teams • Workshops to develop innovation capabilities • Collection of ideas in organizations • Innovation camps • Open innovation initiatives	• Design thinking projects • Design thinking labs • Incubators • Stronger structuring and measurement of the innovation process	• Transformation of the organization through design thinking • Application of entrepreneurial thinking in management (so-called effectuation)

Innovation camps: Innovation camps are short, compact workshops that can last up to a week. They can be equated with education and training, but they produce more concrete, tangible and actionable results due to the longer time allotted to working on question statements. During innovation camps, multiple teams typically work on either the same or different question statements that are relevant to the organization.

Spaces: Due to increased understanding of design thinking, an increasing number of design thinking labs and ad-hoc spaces are being set up.

Level 3 — Catch-up

Organizations at level 3 view design thinking as a valid, universal approach to solving numerous problems and challenges within the organization. In line with this is the use of design thinking in all its operational forms within numerous parts of the organization.

Spaces: Design thinking requires space in different forms. Organizations at level 3 typically designate design thinking floors to design thinking projects.

Structure and measurability: The advancement of design thinking implementation in the organization is evaluated, with attention paid to measureable criteria that are permanently followed.

Internal incubators: Internal incubators — setups in which employees work only on innovation — are organized. Design thinking is an established tool for working on projects in this context.

Level 4 — Design and Customer-oriented Innovation

Organizations at level 4 have completely taken on and applied the design thinking approach throughout the entire organization. Alongside the operational processes, strategy within the organization is completely user-oriented and focused on innovation:

Management principles: Traditional decision-making mechanisms based upon causal logic are complemented by founders' approaches to management. These are hidden behind "effectuation", a decision-making logic that is particularly useful in situations with a high degree of uncertainty.

Transformation by design: In order to ensure the continual adaptation of the organization to the world around it, design thinking is applied to the transformation process itself. Through this, processes within the organization are analyzed, realigned or optimized using design thinking. In this context, design thinking is understood to be a tool for transformation.

RECOMMENDED ACTIONS AND SUCCESS FACTORS FOR THE IMPLEMENTATION OF DESIGN THINKING

The successful introduction and anchoring of design thinking in an organization requires a lot of stamina and patience. A fundamental change in how employees think and work is not something that can be expected within a few weeks or months. It usually requires years. Based upon the numerous projects we have carried out, both in consulting and at the University of St. Gallen, we have identified a few success factors that lend themselves to this journey:

Customer and user orientation

The most important recommendation for action is to focus the organization upon customers and users of products, services and business models, starting from strategy through to single activities within processes. Above all, an organization's strategy determines the course of the management's activities, and relatedly, the systems of goal setting and measurement for employees. Because of this, the company strategy should include steps for implementing and achieving customer and user orientation as early as possible.

Management commitment

Unfortunately this term is overused, but nonetheless the success factor of management commitment has its value. If the support of upper and middle management fails to focus the organization on customer and user orientation and innovation, or if what they say is seen as mere lip service without concrete support for employees, then the introduction of design thinking is possible only with enormous difficulty. This support is necessary, as design thinking breaks the learned behaviors of an organization and, at times, transforms them into their opposite.

Without the backing of management, many steps of the process — such as, for example, the divergent phase — cannot be integrated into the classic employee incentive system. Besides this, there is often substantial resistance from many employees. Therefore, managers in organizations must foster the right conditions for design thinking to be able to grow.

Lighthouse projects

Nothing convinces people more effectively than results, so the successful implementation of design thinking is often demonstrated via lighthouse projects. Instead of analyzing the potential positive effects of implementing design thinking with colleagues at an abstract level, lighthouse projects quickly show the value of design thinking for the organization. Typically, problems are chosen that can be manageably resolved within a period of six months or less. These results often spur the initiation of change.

Sharpening perception and understanding

Parallel to lighthouse projects, training and educating all levels of management in organizations has proved useful. Training and education help design thinking become a part of the organizational culture; they establish the foundational knowledge and understanding of design thinking methods. Successful organizations focus on training many employees rather than just a few promoters.

Selection and focus

Design thinking is not a method that can solve all the world's problems. Define the concrete uses and parameters of the implementation field that makes sense for your organization. In later stages of implementation, the parameters of use can be revisited and in some cases revised.

Stamina

Change in organizations is a long and difficult journey. Employees must recognize the need for change if they are to carry out learning processes that address the need for change. Change and learning processes require much time. To find this time — usually years — consistency and stamina within the organization are necessary. Unfortunately, in many cases of organizational change, the necessary concentration seldom endures. Nevertheless, stamina across all levels of management is necessary to mastering this transformation.

Break up silo structures

A further success factor for the introduction of design thinking into organizations is the break-up of silo structures. Silo structures partition the organization in terms of competence areas, between which only formal communication channels allow for the exchange of information. For design thinking, these formal channels are restrictive. Instead, interdisciplinary teams that can freely communicate without the consideration of organizational silos are required.

Small teams

The success of design thinking projects is not typically defined by team size. Experience shows that small teams can have an advantage over larger teams when it comes to agility, creativity and a "do it" mentality. Therefore, when possible, these teams should be given priority.

Spaces

The preparation of adequate spaces in which teams can work and also leave their materials undisturbed is essential. Therefore, create the right work atmosphere. Ad-hoc design thinking spaces are already enormously helpful for teams.

5

CASE STUDIES

These case studies use examples from organizations to show implementations of design thinking and the challenges associated with them.

CASE STUDIES

DEUTSCHE BANK

ABOUT THE AUTHOR: KATHARINA BERGER

Katharina Berger has been an employee of Deutsche Bank since 1978, and has worked in very different departments. After starting with customer consulting, her tasks included developing the incorporation and application of PCs within branches. The rollout also fell within her responsibilities, which included consolidating server landscapes and establishing user support through a helpdesk. Through her work on international projects and global test management, Berger built substantial experience in intercultural teams. She has been the head of design thinking in innovation management at Deutsche Bank since 2008.

Short description of the company

Deutsche Bank is a leading, global, client-oriented universal bank with 28 million customers worldwide. It offers clients diverse financial services, from monetary transactions and credit, to investment advisory services and asset management, to all forms of business in the capital market. Its clients include private individuals, mid-sized companies, public funds and institutional investors. Deutsche Bank is Germany's biggest bank. It has a strong position in the European market, and a significant hold in markets in the Americas and the Asia-Pacific region.

What were the catalysts for starting to use design thinking?

On the way from the information age to the age of the customer, customers and their needs must be positioned more firmly in focus than before. The speed of technological change and megatrends such as demographic change, digitization and the "internet of things" have, in addition to many other megatrends, starkly changed the lives and needs of our customers. Design thinking as a method for developing new ideas has the unique charm of enabling employees to explore the needs of the customer from the very beginning. They can systematically research the possible solution space, and challenge the original problem professionally. This enables the successful adaptation of products and services, even in an environment that is changing very quickly. When we started with the first prototypical run-through of a design thinking project in 2008, the following catalysts were decisive for our first "try":

Customer orientation: Design thinking is foundationally conceptualized as a customer- and user-focused method. Through this, we saw the opportunity to enable our product and software developers across departments, and also in Group Technology & Operations (GTO), which is focused on infrastructure, to better understand the requirements and needs of users in relation to new and also radical products. This gave us the opportunity to compare our assumptions with the reality of the client, thereby reducing risk that we would miss needs in development.

In this approach we also saw a way to initiate communication between the technology department and business units earlier. This connection in the early phases of the projects led to an improvement in internal communication, enabling us to incorporate available expertise at the start of projects, integrating the customer early and actively in the innovation process.

Prototyping and agility: The connection of short, iterative processes with the constant building of prototypes proved to be advantageous, allowing us to address market requirements quickly and with agility. The quick iterations, in the form of feedback from users, enabled us to dynamically further develop our ideas and concepts. Similarly, we procured a mechanism through which we could identify false assumptions about product and service design early in a given project through user testing, which helped us reduce project risks.

Innovation and divergent thinking: Thinking outside the box was a deciding factor in trying out this method. Design thinking connects the search for alternatives, at times radical, innovative solutions, and specific, concrete requirements with development projects in a unique way. It fosters free thinking. The approach of prototyping early with simple materials allows the project team to evaluate a large number of user ideas. Its methods enable us to deepen our understanding of the problem and user needs, and extend our idea space. If consistently applied, they enable the project team to consider a given issue from various perspectives. What's more, design thinking fosters this perspective-changing intensely. Through this, we can position our choices in regards to a given approach on a significantly stronger foundation. Design thinking is substantially more than just a method for creativity. It is an outstanding supplement to agile development approaches in relation to creativity and innovation.

We conceptualized a first approach in regard to the use of design thinking. As a result, in Eschborn, in Frankfurt, a bank-focused design thinking laboratory was founded, the idea factory. Subsequently, the lab, equipped with simple materials, grew into a strongly established space for ideas.

How do you define design thinking?
For Deutsche Bank, design thinking is one of many methods helping us work on customer-oriented and agile new products and services. It is aligned with our organizational strategy, and places our customers at the center of everything we do. In the past years we were able to turn design thinking from an experimental method into a mature approach adapted to the banking sector. Related to this, Deutsche Bank implemented the following initiatives:

Projects: Since 2009, we've carried out successful projects with customers and also with suppliers. A meaningful element was that Deutsche Bank decided early not only to include internal employees in projects, but also to offer university students from all over the world the opportunity to take part in these projects through internships. Through the focused networking of young talent with internal experts, defined as a goal in the project setup, we were able to incorporate new perspectives on the problems we face. Furthermore, we were able to offer our employees the opportunity to experience the approach early. As a result, we were able to also take a step in fostering change management. In the last few years, our projects are focused not only directly on our customers, but we also use them to work through internal questions for the betterment of our own processes and information systems.

Training: The design thinking lab has, in collaboration with the University of St. Gallen, actively supported projects and trained employees worldwide. At the Deutsche Bank headquarters in Frankfurt, London, New York and Singapore, over 300 management trainees were schooled in the method. The goal is that as much internal knowledge as possible is identified and collected, flowing directly into operational work.

Two years ago we began to integrate elements of design thinking into classic projects in the GTO department.

What happened with your project results?
The large number of results generated in our projects flowed in various forms into new products and services. We developed new solutions such as the future planner, an application that enables our customers to put their wishes on a time axis in a financial context. On the basis of the customer, not offers from the bank, we have a dialog with our clients.

What's interesting is that our projects have had other noteworthy effects. Let's take a topic like big data. Right now it's being discussed intensely in numerous organizations. If one approaches it through design thinking methods, one arrives quickly at the aspects beneath the surface: Which new roles and abilities do we need if we want to address this issue? How can we enable uses, also in smaller departments of the organization, when we do not want to build up a team of experts? How can we achieve an exemplary approach to making big data understandable, and enabling it to be experienced? Thanks to our design thinking team, connecting experts and potential users was intensified. What's worth paying attention to was that through the team working with us, a perspective in the expert discussion was picked up that had previously been missing. Through the team's approach, the very technical discussion was transferred in a holistic way. This is a valuable effect in the early identification of risks and achieving unexpected solutions.

What have you learned through design thinking and about design thinking up to this point?
The design thinking method has taught us many things:

Customers: It's always worth it to involve the customer directly in the process. Insights from individual conversations and observations can be directly implemented in new products and services.

Value: When an organization wants to strengthen innovation, it must often also achieve a new mentality; that researching problem and solution spaces, as well as direct exchanges with customers, are a central component of professional work in projects. Design thinking helps us with its toolkit, but also with the approach of integrating and testing ideas and concepts through an iterative process. Through this, employees are given a tested way to achieve change.

Implementation: The successful anchoring within the organization takes time, lots of time! This time is also necessary to initiate and carry out the change process sustainably. One must choose whether one wants to achieve a cell, or an innovation island, or bring the whole organization on the journey. Time is also important when adapting the methods of design thinking themselves to the organization. Many steps are required. The design thinking team at Deutsche Bank had first of all to understand the methods, and then develop an approach regarding how to use them directly in projects. As a further step, they had to find a way to allow the methods to flow throughout the whole organization. What's important here is that the methods are in very few projects applicable in their pure form. The integration in agile development methods still offered successful approaches and solutions.

How do you want to continue using design thinking?
Design thinking definitely remains an elementary component of our portfolio of methods, also for future projects. What's important is that the methods find their way into stronger operational use, for example, in the way that elements are integrated more intensely in other approaches, such as agile development methods. Similarly, we will continue to review which aspects of design thinking offer a unique competence in which area of the organization. An example of this is the topic of customer journeys, and relatedly, personas. These methods have been used for a

long time in marketing, but how can we connect a competence such as this directly into the innovation process and the operational concept? We will address these and other questions in the future, supporting further integration.

What advice would you give to other organizations?

Design thinking does not run itself. Generally, cultural differences between daily life in a classic management context and the requirements of design thinking can be felt. Most of us were raised to solve problems as quickly as possible. However, through this, it's easy to tread the path already taken, missing out on new ideas. Design thinking is consciously iterative. Calling into question one's own assumptions and allowing oneself to take this approach is not easy to impart if one has never been part of a design thinking process. Therefore, here's my advice: Involve the decision-makers in the journey from the beginning. Tenacity, endurance and consistent implementation are necessary for the realization of design thinking. Of course, in pilot projects, fast results can and must be shown. But these cannot dissuade one from the fact that complete uptake and implementation of design thinking as a part of the organizational culture takes time.

Furthermore, professional support is indispensable, right at the beginning. For an introduction, deep expertise is required, even when it's tempting to think, and is often even claimed, that the methods can be easily used with a simple basic understanding. It's certainly the right choice to empower employees to gain experience in early projects and use of methods. But we cannot fool them regarding the path to complete implementation. This can only result from a basis of expertise, through experts. To achieve a mature concept and to take advantage of the full value of the methods, new experts and new skills in the organization are required. In order to get there, it's highly advisable to search for external support until the skills are built internally. The selection of experts should ask whether the person in question has experience introducing the method in an organization, not just with the methods themselves. Over time, one will need this expertise in organizations. The investment in employees is an investment in the future of the organization and ability to offer customers the products and services they want.

CASE STUDIES

SWISSCOM

ABOUT THE AUTHOR:
FRANK SEIFERT

Frank Seifert has worked in various management positions for Swisscom since 1999, and since 2010 has supported the development of human-centered design (HCD). He developed the internal approach to consulting and led it with private clients, IT, Network and Innovation and Group Security. He has led Swisscom's external HCD consulting team since 2014.

Short description of the organization
Swisscom is the leading telecommunications company in Switzerland, with its headquarters in Ittigen, near the capital of Bern. Over 21,000 employees have achieved sales of 11.7 billion Swiss francs. To customers, Swisscom positions itself as the best companion in the connected world, where customers can be assured that the focus is on both the essentials and discovering new opportunities. Swisscom offers its customers the best in the connected world, always and everywhere. Swisscom offers business and private customers a cellular telephone network, fixed lines, internet and digital television. It is also one of the largest providers of IT services in Switzerland. It handles the building and maintenance of the cellular and fixed-line infrastructure, distributes broadcasting signals, and is active in the fields of energy and health.

The Swiss telecommunications market has an estimated annual sales value of around 13 billion Swiss francs. Depending on the category, market share fluctuates between one- to three-fifths in each category. Telecommunications and IT have been merging into a single market for years. The market is increasingly bound by the rules and laws found in the IT sector: high speed of change, standardization through market power (not standardization committees), quality models from IT and global business models. This creates a unique situation for Swisscom because while the company Swisscom works in a regulated national market, it is simultaneously dealing with international competition that also position itself as a direct partner for customers, curtailing Swisscom's ties to customers and its sales volume.

What were the catalysts for starting with design thinking?
Until 2008, Swisscom was a telecommunication organization organized according to technology. Fixed lines, cellular coverage and internet were offered and provided by different departments. The result was that customers did not have anything to do with Swisscom, but had to address whichever area of operations was responsible for their concern — for example, Swisscom Mobile. In part, the competition was coming from new organizations that were active worldwide, such as Google, Facebook and Apple. Because of their internet technology and business models, they wanted to take over an essential piece of the market. This competition was omnipresent.

So that we could more strongly differentiate ourselves and work in a more customer-oriented way in this environment, our concerns were reorganized according to three main groups: private clients, small and medium-sized enterprises (SMEs), and large clients. At the same time, we wanted to take a big step and develop our offers in a customer-oriented way. Instead of classic project management, we changed to the customer experience design (CED) approach — this against the backdrop of a Swiss communication market in which Swisscom has a market share of over 50% as well as a price premium — which, according to economic theory, is very difficult to achieve. The theory behind CED (see Pine & Gilmore, 1999) suggests that customers are prepared to pay higher prices for better experiences, and are simultaneously loyal. The brand also comprises an important element. CED focuses on experience, thereby strengthening the differentiation of the brand not only in relation to its direct competition, but also brands in other markets, as the effect of the experience is evaluated across market boundaries.

However, a product manager cannot become a customer experience designer overnight. The question that quickly arose was, what are the key characteristics of a customer experience designer? What makes him or her different? What methods does he or she use? At this time, Swisscom was able to profit from the fact that a Swisscom employee had worked for years at a Swisscom outpost in Silicon Valley with Christina Taylor, and had explored what made the start-ups and other firms there successful. Two essential elements were the keys to success: Customer journeys and design thinking. Both approaches demonstrated the basis on which to develop the appropriate methods and mindset for Swisscom. These became the foundation for human-centered design and training principles for new customer experience designers.

How do you define design thinking?

With the approach of turning product management into CED, the question of what CED is quickly arises. The important components are a mindset and a method set. The mindset can be understood as a tool. As a catalog of questions, leadership helps define the context, allowing a culture of innovation to develop and results to be achieved. The method set includes a range of methods that helps when developing offers. The essential elements of this method set, based on design thinking, are:

- It is not the customer's job to say which offer they will use in the future. This is the work of the customer experience designer. He or she will only manage this when he or she observes the customers' needs as directly as possible. Reading a market research report does not develop empathy for the customer.
- To find the most successful path when designing customer experience, one must test single approaches to solutions with customers using prototypes. Through this, one quickly and cheaply learns which offers should be created.
- A third important element of the method set for us is, the customer experience chain. This describes the essential elements through which the customer experiences an offer. It consciously helps create the elements with which one would want to influence the customer experience.

In the first phase, more than 600 employees and managers in the private customer area were trained in the mindset and method set. The training itself was a multi-day experience which bound theory together with actual usage. In the second phase, a pool of internal consultants was assembled. These ensured that the mindset and method set came into use within selected projects. It was important to build this consultant pool up from our own employees. Through this it was guaranteed that the consultants — called CoCreators and Validators — took on responsibility for the project results and the use of methods as members of the project team.

As a consequence, the first successes could be shown, and they encouraged the organization to continue using the approach. Prompted by this confirmation, a range of different projects employed design thinking in an effort to gauge the universal applicability and/or limits of the approach. A critical factor is that the mindset and method set supported the existing development processes for offers, but did not replace them. It is important to mention that both these roles were supported by experts in the fields of market and trend research, communication, visualization, prototyping and work environment design.

In the third phase, offering the best customer experience became part of our strategy. This approach, previously developed and used above all in the private client area, was broadened to cover all customers. To achieve this, additional teams of consultants for the SME and large client areas were established. Consultants were also made available for internal departments such as human resources, IT, network and innovation, group security and marketing. In the fourth phase, we established consulting teams for Swisscom's large clients. In parallel, over years we built up a pool of facilitators. Facilitators are employees that are trained specifically in the mindset and method set, so that they can apply them in projects both within and outside their typical work areas. When facilitating, a superior relieves them of their usual tasks for a designated period of time. Through this, the operating area of the consulting teams substantially increases.

What happened with the results of your projects?

The results we were able to achieve have confirmed the validity of the approach. The conscious creation of experiences aided by design thinking has become an established competence at Swisscom. Impressive examples of success are the newly developed Swisscom Shops, the package offers for internet, fixed lines, digital television and mobile coverage, Swisscom TV 2.0 and the new pricing plans for mobile telephone calls for private and business customers. These successes would not have been possible if numerous internal obstacles had not been tenaciously taken on. Even the name — "design thinking" — can lead to misunderstanding and dismissive reactions. In German, "design" implies the creation of forms and representations. Because of this, design is often placed at the end of a development process to make technical innovations "attractive". Design thinking, however, is effective only when it is consistently used, particularly at the beginning.

The term "thinking" also leads to confusion. Design thinking methods are extremely action-oriented and leave less room for boring, time-consuming discussions. Design thinking is not a method that can be taught to employees "only" to bring better results. If one wants to successfully carry out design thinking, management must also be ready to take on another management style. At the beginning, design thinking will often be perceived as time-intensive, and a slower process. In a similar vein, it was said that when using classic project management methodology, a short analysis of the actual problem statement followed by solution finding and a boomerang into the same in later project phases kicks back and makes projects tiresome, unsuccessful and expensive. The introduction of design thinking is therefore a fundamental change — sometimes a radical change— and this requires time. In our experience, these changes lead to resistance in the organization, as we know from change management.

What have you learned through and about design thinking up to this point?
Seven years of experience in the introduction and usage of design thinking has taught us a lot:

- Not every project in which design thinking methods are used must lead to success. Design thinking is not magic. But it greatly helps us to ask the right questions at the right times, and put a stop to developments in the wrong directions.
- For design thinking to be successful, the project team must be sufficiently heterogeneous. This means different perspectives, know-how and thinking preferences, for example in line with Hermann Brain Dominance Instrument (Hermann International, 2014). This influences how the team approaches challenges. A good use of design thinking methods does not make a poorly assembled team a success.
- New employees do not come into the organization bringing design thinking know-how with them. That means there is a permanent need to train new employees.
- It has been worthwhile not to view design thinking as external knowledge that is brought in when a project needs it, but as a competence that one builds with one's own resources. Through this, design thinking has become a standard method in many areas of the organization.

How do you want to continue using design thinking?
For us, design thinking is a part of our HCD approach. Since we have the goal of establishing HCD as a core competency of Swisscom, we will continue using design thinking. Over the past few years, we have increased the size of the HCD team to 60. These employees support various teams throughout Swisscom's business divisions. The staggered roll-out has resulted in varied HCD development statuses at Swisscom. The goal is to achieve a consistent stage of development over the next few years.

What recommendations do you have for other organizations?

Looking back, the following measures were critical to success:

- It takes long-term support from top management and a strong team with strong personalities in middle management. Success is very dependent on the activities of the people involved.
- At the beginning, we formulated a clear "why" for the introduction of HCD. The why was focused on the Swisscom brand, and its ability to differentiate itself. Relatedly, there was a clear business relevance, and could be concretely measured using a net promoter score. The ongoing digital transformation once again confirmed our clear orientation towards people and the customer experience. Customers have never before had the power they have today. The sales channel is a screen; information and prices are completely transparent to customers. Only offers that quickly lead to positive experiences without obstacles are recommended by customers over social networks. These secure the company's success.
- We chose to take a holistic approach: A clear why, an effective strategy, the right structure and organization, and measurement criteria that show us how we are doing, the right people and methods, as well as cultural elements such as specific spaces created for HCD. It is better to take a holistic approach in one limited area of the organization than to use pieces across the whole organization.
- The why and the implementation were promoted by an area of the business with direct effects on business results. Relatedly, we have differentiated ourselves from classic approaches to organizational development, which tend to be strongly rooted in human resources. As a result, other business divisions see these other approaches as less critical to success, and are less likely to support them.
- Nothing is sexier than success. It is important to deliver examples of successful projects as quickly as possible. These add momentum to the change process. Selecting the appropriate projects for this is not always easy.
- The selection of a method set that fits the organization and its scope is important. Over time, one can observe which methods are accepted, which terms are incorporated into the organizational language, and how they are discussed in decision-making.

CASE STUDIES

HAUFE-LEXWARE

ABOUT THE AUTHOR: CHRISTIAN STEIGER

Christian Steiger, born 1973, is the director of business operations at Haufe-Lexware GmbH & Co. KG. In this role, he is responsible for the business division that delivers new products, business models and innovation. He is also responsible for the building and further development of SaaS EcoSystems lexoffice, developing online solutions for small and medium-sized enterprises, as well as the establishment of new technologies, methods and procedures (Wave strategy, agile development and design thinking).

The Haufe group is a digital media group. We offer digital solutions to optimally set up workspaces for all the target groups in our solution portfolio. Under the brand name lexware and online (connected) solutions, users can bring order to both their business and private finances, from book-keeping and merchandising to taxes. Haufe provides electronic products and online offers, commercial user software and books as well as periodicals on law, economics and taxes.

With the Haufe Academy, our customers' qualifications are available from a single point, in the form of open, bookable events as well as individual, in-house seminars, customers receive offers for customized, supplementary training and development.

In order to keep up with rapid changes in both markets and technologies, it is critical to find new fields in which to become active in business, and become economically successful within them. In 2010, we established a new, completely independent department called "Blue Ocean", based on blue ocean strategy. The goal and philosophy of the department is to implement blue ocean strategy in action. Blue ocean strategy states that only relevant, differentiated needs can lead to new markets, and thereby new clients and long-term success. To put it simply: in the future, good money will be earned only in markets where there is little or no competition. The book "Blue Ocean Strategy" (Kim & Mauborgne, 2005), describes methods still used today, such as strategic contour and ERSK — *Eliminierung* (elimination), *Reduzierung* (reduction), *Steigerung* (increase), and *Kreierung* (creation). It also describes companies that have done things differently than their competition. However the question of how the firms discussed came to the new issues and needs of their customers and entered these new markets free of competition generally remains open.

We have developed the strategy "Wave (scout, proof, deliver)" to find new markets. The first phase ("scout") is the need-finding phase, and plays a central role. The "proof" phase follows, in which we carry out business modeling. Finally, "deliver", is the final product concept. In the "scout" phase, we implemented design thinking as a method. Here, we have the goal of using need finding to solve complex problems with innovative ideas. The principal approach in design thinking is similar to our agile organizational approach: new ideas only arise in an interdisciplinary team, which together develops a question that has peoples' needs and their motivations as a fundamental point of departure. The concepts are then implemented and reviewed from multiple perspectives (such as lean management, agile organization, agile software development [Pautsch & Steininger, 2014]). This iterative and agile (in German *"beweglich"*) approach is in the DNA of our department, and anchored within all processes and phases of the Wave strategy.

In the context of our Wave strategy, we decided to pilot a design thinking project. Initially, we formulated a visionary challenge with a problem statement in order to provide the future team a rough frame. The project started with a briefing, and having the right question was fundamental. We worked on this question internally within an interdisciplinary team, and then finalized it with Falk Uebernickel and the design thinking teaching team at the University of St. Gallen.

The goal was to generate prototypes of approaches to solutions that take online activities into account. This included the proposal process, winning customers, customer relations, service provision, back office and administration. With a focus on these, the solutions proposed should show how management and related organizational challenges might look for small and medium-sized enterprises in the future (within a five-year maximum time horizon). The focus was predominantly on those in service provision, freelancers (excluding employed persons without reporting requirements), craftsmen and others, as the team sees relevant.

We carried out the handover and briefing on-site with the team, and assigned an employee as the central point of communication. The team was comprised of three master's students at the University of St. Gallen (HSG), Haufe-Lexware employees, and instructors at the University of St. Gallen and Stanford. During the kick-off event at Stanford University, we had the opportunity to be on site at the d.school during the presentations and workshops on methods and philosophy by Larry Leifer, among others. In addition to well-known methods (such as the empathy map and personas), we also got to know the Business Model Canvas (BMC), which is today a central component of the aforementioned Wave strategy during the proof phase. Here, the BMC summarizes the results of our business modeling.

The design thinking project had a timebox of five months and was divided into five phases. In the first phase, design space exploration, the goal was to observe people facing a certain problem. Here, we could learn how quickly we try to interpret and come up with solutions, instead of observing and listening. For us, the methods of explorative interviews, personas, empathy maps and question funnels were particularly interesting. We continue to use them today, dependent upon the context and question statement. Then, we continued with the critical function prototype (CFP), with the goal of identifying the critical functions that could lead to a solution from the perspectives of users. The methods we used were prototyping (scribbles, PowerPoint screens and also "thrown-together prototypes" such as a "ring with scanner"), where we included a particular critical function and tested it with users.

The dark horse prototype phase (DHP) was particularly fun, since it had the goal of turning all of our assumptions and experiences up until that point on their heads, to help in the search for innovative solutions. Here, we used methods such as "stretch the goals" — that is, pushing the goals so far that current solutions are no longer sufficient. Two examples are the "solution for tax evasion" and "do the pig", which is a role-play activity involving all stakeholders, including the client, the tax consultant and the financial authorities.

We also carried out framing and reframing. With framing, the issue is identified and the basic assumptions are discussed. With reframing, these basic assumptions are reversed. In the subsequent funky prototype phase, the goal was to refocus and reorganize needs. Relatedly, the ideas for solutions were newly tested and further developed, as informed by successfully tested functions from the CFP and DHP phases. The "wish list" offered a picture of the functions that the users had liked the best up until that point. Through the idea of "let others do", people external to the project were brought in to develop new approaches to solutions. Members of other teams did this. In the functional and final prototype phases, the final prototype had both horizontal and vertical elements. For us, this was a completely web-based prototype that included all the successfully tested functions identified in the preceding phases.

Besides the final prototype "Somba" — the primary result — we received all the partial results of the individual phases in a 250-page final document.

Somba stands for "social micro business administration", but is also the name of an African people who organize themselves using similar structures to small businesses.

The user promise of Somba:

- Somba is a web-based workspace that is always accessible.
- Somba automatically logs administrative background data.
- Somba allows for the improved exchange of information between business partners.
- Somba is paperless.
- Somba is focused on people, and therefore fits into the structure of small businesses.

These results have directly flowed into the conceptualization and realization of our lexware solutions today. Lexoffice is an invoicing and book-keeping program for small businesses, freelancers and the self-employed (see www.lexoffice.de). The feedback and customer satisfaction with this solution show the value that the design thinking approach can add to real products.

From our perspective, design thinking is of interest to all employees of the Haufe group whose work involves conceptualizing new products, solutions, or services. Design thinking, however, can only be seen as part of our Wave strategy process (see the Wave scout phase). We intensively use design thinking in the scout phase. After Somba, we realized two more design thinking projects with the University of St. Gallen, and certainly will again. Within the Haufe group, we started additional projects in the field of design thinking with support of the University of St. Gallen and others.

In our experience, it is important to direct a strong focus towards need finding and prototyping, and to clearly separate business modeling (feasibility, business model, etc.) from these activities. However, in business modeling language, design thinking delivers two of the nine essential building blocks, namely the value proposition and customer segments. For us, the other building blocks are completed in the subsequent proof phase of business modeling.

CASE STUDIES

MARIGIN

ABOUT MARIGIN

Technologically cutting-edge and both conceptually and culturally innovative, "Marigin — Center for Animal Medicine" is a newly founded veterinary hospital in Feusisberg, Switzerland.

The organization is based on a very successful veterinary clinic that was built up with the help of private investors.

Marigin offers a full range of services, from modern diagnostics (laboratory, digital x-ray, CT, ultrasound and endoscopy) to special therapies and operations by selected specialists, to physiotherapy with a water treadmill and alternative medicine. Supplementary services such as a veterinary shop and cat hotel round out the offering.

Marigin is a medical one-stop experience for pet owners.

How did design thinking flow into this project?
Initially, it was more traditionally planned: functional, treatment-specific procedures would be analyzed, conceptualized and optimized by experts, and the architecture would be organized accordingly. A rational use for the building was developed, that in addition to veterinary functions, offered tenant's space.

The team responsible, however, had the feeling that this purely rational-functional approach did not reflect the full potential of the operator, and formulated the wish for a central theme that emotionally connected all the functional aspects in an appreciable fashion.

For this, they searched for an external innovation partner that would help them to iteratively elaborate these emotional aspects together with all relevant stakeholders. The design thinking process was initiated.

Who was involved?
The design thinking team consisted of the leading doctors, assistants, investors, customers, architects and some lateral thinkers foreign to the subject. New people were always integrated into the process for given periods of time to prevent the perspectives from entering a static loop.

What was the concrete procedure?
First, the existing procedures and needs of the different user groups were analyzed in terms of their precise strengths and weaknesses with the help of persona and touch-point analyses in relation to existing practices. Then we generated an optimization profile.

The process knowledge gained was useful as an affirmation, and bolstered basic quality, but it was not surprising.

Planning mistakes that had already been made were quickly identified. For example, the original plan called for a large, traditional reception counter. This implied a barrier from behind so that the assistants would not even be able to see their four-legged patients.

Really high potential for innovation emerged on emotional and cultural levels.

With this learning, Marigin decided also to use the DNA analysis method (see p. 82), in order to identify the essential core of the organization. The goal was to define these existing core characteristics, and secure and scale them for the future.

The team intuitively felt that this specific emotional core was the real reason for the enormous success the organization had enjoyed up until that point.

What was the most relevant outcome of the DNA analysis?

Although the practice already had an extremely high number of clients, it was possible for the staff to remember names and personal details, and to use these during interactions with clients and their animal patients.

Additionally, casual conversation with one another and with customers was unfamiliarly friendly and direct — very open, and marked with real sympathy and credible interest.

The whole staff differentiated themselves through their attractive and approachable dynamic, which broke through the common "white standard" effect of a traditional practice in a pleasant way. All of this was possible without giving an impression of diminished seriousness or professionalism.

Ultimately, this unconventional cocktail led to a regular, widespread fan club that extended far beyond the normal preference for a doctor.

How was design thinking implemented?

It was now about extending these unique emotional aspects that initially appear coupled to a limited size of organization to one of the largest clinics in Switzerland. The recipe for success should of course not be lost.

The design thinking team described previously began a completely free collection of ideas regarding how this individuality could also be achieved on a larger scale.

Being able to greet clients personally was an important aspect, but with 2,000 customers appeared to only be partially possible and certainly not with 10,000. During brainstorming, the following ideas came up: automatic patient recognition upon entering the practice through a chip implanted in the animal, and automatic availability of patient data before interacting with the customer and patient, including an index of existing profiles (including shopping lists, pet food preferences, etc.).

In the process, ideas were always quickly sketched, and sometimes readjusted to gain feedback as quickly as possible. On the basis of these reactions, the ideas were revised, presented again, and then once again critically reviewed until the team had the right solution in terms of content, and could hand it over to the realization reality-check.

Another idea was the desire to build up personal interaction, and reduce "impersonal" interaction. This led, through the aforementioned design thinking process, to the idea of building of a costly, medicine-distribution robot. In the future, this would prevent staff from having to walk away to retrieve medicine, allowing personal interaction to carry on uninterrupted and eliminating waiting time.

Primarily using design thinking together with clients, the idea was developed that instead of a standard, bordered waiting room, there should be areas that made communication easier between pet owners, since they are typically so interactive with one another anyway. Through this, bar tables to foster communication were developed. These also serve to establish separate zones for dogs, for example. The inner space was implemented with the same qualities as the outside waiting areas, as animals tend to be much less nervous waiting outside than inside.

Another factor was the unpleasant smell that normally permeates veterinary hospitals: Special air purifying systems, a spring and aromatic herbs were found to be solutions during the design thinking process, and were implemented in reality.

In the clinic, a room for dying was integrated in a sensitive manner, which delicately took into consideration the strong bond between the human and animal, and the uniqueness of the situation.

What was unique in the process?

The combination of different methods (DNA analysis, personas, touch points and design thinking) quickly led to an emotional and strategic "panoramic view" that easily brought the potential to light.

Above all the design thinking process, with its quick interactions around visualization and improvement, helped realize solutions efficiently. That was critical from the beginning, due to a tight schedule.

It was about developing realistic, innovative landscapes and being able to quickly reach decisions. It was explicitly not about developing unrealistic visions and time-wasting *PowerPoint* presentations.

The design thinking process proved itself to be the right approach in this case, and led to innovative solutions that were able to be implemented.

What do we need to pay attention to?

Since the decision and realization teams were firmly stuck in the operational scope from the very beginning, it would not have been possible to reach such results without the help of fresh professional coaching through external partners.

These partners also helped when it came to the concrete realization of ideas, which was of course more expensive because of the numerous innovations. When choosing partners, one should also ensure that they are not only capable of generating strategy papers and rough models, but can also handle their concrete implementation.

How will this design thinking attitude continue in the organization?

Firstly, there is already a large surplus of innovations and ideas from the design thinking processes we have carried out to this point. We will be busy implementing these for a while.

Secondly, check-ups (à la "mystery shopping") will be regularly carried out, so that the goals and ideas can be reviewed, allowing new tendencies and learning to be integrated.

The process is therefore not a one-off thing that ends, but is rather hidden within a philosophy that promotes optimization and innovation potential being proactively and continuously searched for within the organization.

© Marigin AG, milani design & consulting AG

CASE STUDIES

MEDELA

In a company history spanning more than 50 years, Medela has grown from a small, predominantly technology-focused company in Switzerland to a global, family-run producer of the most advanced milk pumps and medical nursing systems. From the beginning, Medela developed research-based, safe and efficient products and thereby always set new standards. In addition to other approaches, design thinking was introduced within the organization to bolster its position in the market.

Medela is the global market leader for breastfeeding products. Breastfeeding is indisputably best for the baby. Medela is the godparent, and upon the basis of scientifically proven fact, offers various products and services for infant nutrition worldwide. This focus is at the center of all activities of the Medela mother's milk sector.

If mothers cannot or do not wish to nurse directly, or are at times unable to nurse directly, milk can be pumped from the breast to feed the baby. This can be the case when mothers work, or when it is medically not possible to nurse. This is where Medela solutions come in.

What is the innovation culture of the organization?

At the beginning, the company was influenced by the founder and inventor, Olle Larsson, and later his son, Michael. Brilliant technological problem-solving stood in the foreground, and numerous patents and first-market introductions came soon after, and with these, market leadership. Over the decades, the competition grew, along with the pressure to innovate.

For a long period, subsequent innovations were related to technology and mechanical improvements. It was important to establish the design thinking process to serve as a supplementary source of innovation. Although breastfeeding appears to be a very feminine topic, the employees responsible were overwhelmingly male engineers. In 1999, the then director of development integrated a female-run creativity agency, which established the design thinking process in the organization with the support of top management. This step enabled womens' needs, and above all the emotional aspects, to take a stronger place in the foreground, and lead to innovations.

Market research and insights

Since then, we have increasingly focused on user observations, above all in the biggest market, the United States. Through this, we have compiled deep and comprehensive knowledge of our users — mothers — which flow into innovative services, communication, apps and product solutions.

Research

Simultaneously, the founder's son, Michael Larsson, intensively promoted research on lactation at the University of Western Australia.

The result was groundbreaking new knowledge in the field. For example, due to new measurement and observation methods, the natural nursing behavior of the infant's mouth and on the mother's breast was precisely determined and documented. Previously this was accomplished only with a large margin of error.

How does one bring insights, engineering and research together, whether divided in terms of place, content or culture?
Through the neutral support of external experts in combination with the training of an internal team, the design thinking process was established to unify these aspects.

First, we carried out the method of "DNA analysis" (see p. 82), in order to define a collective strategic and emotional orientation. These collaboratively defined, overriding goals laid the groundwork for a homogeneous motivation and relatedly, efficient decision-making processes in the whole team.

In the first step, all relevant content and strategic aspects of Medela were collectively expressed in their most extreme interpretations: How do we act today? Where do we stand? Where are we strong? We then compared all the experienceable factors of the strategic status quo at the time. The deltas between the potential and experienceable contribution were shown and debated.

Based on this, we determined the strategic design goals we wanted to achieve in the future. This DNA analysis process established the necessary pressure to act and the necessary collective understanding for entry into the design thinking process.

How could the process be established?
Through the clear and open support of the owners and chairman of the board of directors, an interdisciplinary process was established. On the basis of the strong collaboration between the various team members from different disciplines such as innovation, design, development, research, and so forth, the next wave of innovation could be initiated.

Who is part of the process?
The lactation scientists from the Australian university bring the newest — otherwise unknown — scientific learning. This is always integrated into concrete solutions.

Internal engineers who enjoy experimenting can also integrate the necessary technical knowledge. Through internal marketing and market research employees, comprehensive insights can be integrated within the team. Mothers and nursing specialists, as well as midwives, allow direct experiences and market knowledge to flow into products. The external lateral thinkers, who have known Medela and its market to be a strong innovation and design partner for many years (but who are also at home in transport, consumer goods and MedTec) also support the process and bring in transfer impulses from other branches.

What is needed for implementation?
Coordinating the process requires the availability of a physical room in which concentrated work can happen, and the interim results can be documented and remain accessible. It must include case-dependent opportunities for international exchange and expert involvement — for example, the Australian experts can participate via Skype.

How does the procedure run?
For example, the results of preliminary research can provoke rough ideas within the team. One idea was to transfer the nursing behavior onto the bottle nipple, since up until that point the nipple was too reactive and allowed too much milk through, which led at times to complaints such as colic. Since transferring research results into practice can sometimes depend upon very complex physical dependencies, in the context of design thinking we use short animation streams to make the basic principles "digestible" for everyone. First models are quickly assembled using alternative materials, which make the functional principles comprehensible for everyone. These are iteratively measured, tested, newly built and further optimized in a ping-pong method. This is a cycle that repeats itself countless times through an iterative process between very different team members.

At the end of this design thinking path, a concrete product solution was developed: the Medela Calma breastmilk feeding bottle. The product solution was then used in the standard process, and has since become well-positioned in the market and won many innovation prizes.

How can the design thinking culture become anchored and visible within an organization?
An approach, from the perspective of design thinking, might be as follows: Employees that are interested must first qualify themselves as team members to push the selection process for the team. Only the best should become part of the team, which substantially boosts its attractiveness and image. Proof of knowledge, out-of-the-box thinking, innovation potential and team spirit is required. When team members are selected, they have the possibility of leaving their existing job profile for the most part, and entering the design thinking lab. Who the design thinking team members are should be openly and actively communicated.

Design thinking and architecture
Everything visible and experienceable in an organization (such as products, services and architecture) mirrors the existing culture. Medela plans to build a central operational and administrative building in Switzerland.

Some of the internal and external team members in the design thinking process have been involved in this planning from the beginning. They are responsible for ensuring that the culture of innovation is given the appropriately central space (a design thinking lab) in this new architecture.

A new build offers the chance to architecturally structure places in a way that makes economic sense, but also inspires creativity, communication, and the strengthening of a living organizational culture. Through the design thinking lab, the motor of the organization should be constantly palpable. Simultaneously, so that the individual teams pull resources into the central design thinking lab, certain floors and departments should also be optically connected with the design thinking lab.

The design thinking lab is more than just image management and corporate appearance; it has an effect that is far-reaching. The central space means that the culture can be felt both inside and outside, and achieves authenticity.

To start innovative processes that need to be further developed and actively discussed, it is advantageous to be in an inspiring environment. This design thinking lab is part of a healthy design thinking culture.

Whoever is offered the opportunity to live and anchor design thinking culture within an organization should use this potential.

medela
Innovation
calma
macht den Unterschied
fait la difference

medela
150
100
50
ml

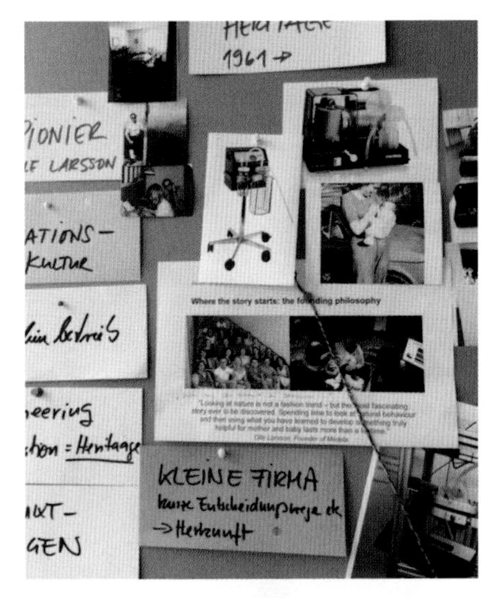
1961 →
PIONIER
Where the story starts: the founding philosophy
KLEINE FIRMA
GEN

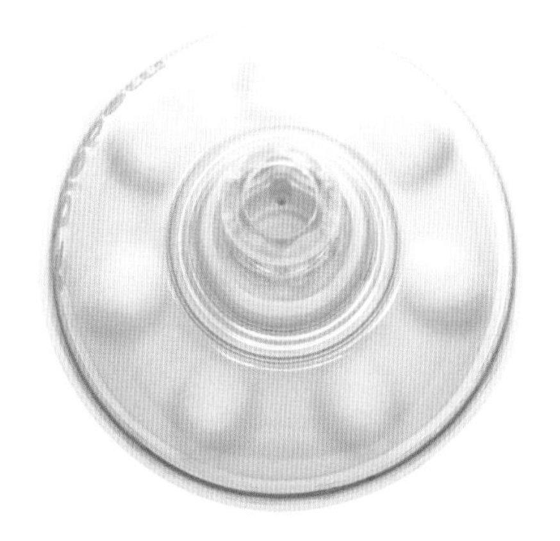

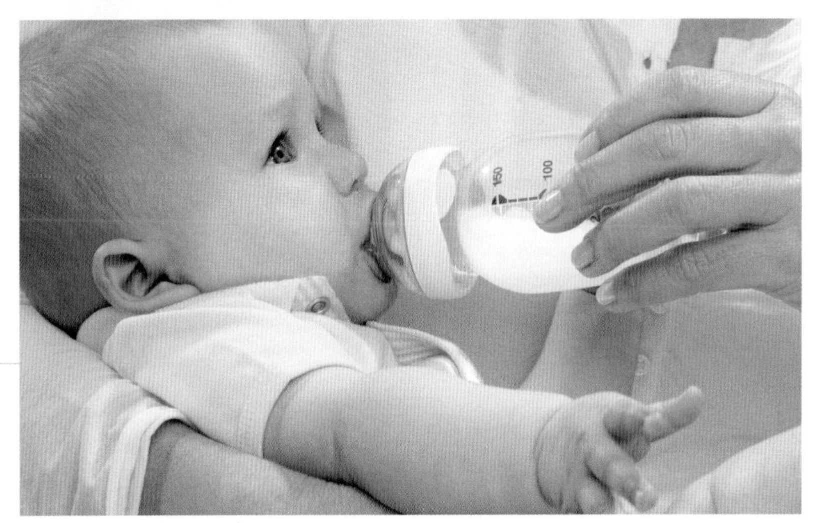
150
100

6

BIBLIOGRAPHY

Allen, T., & Henn, G. (2011). The organization and architecture of innovation: Managing the flow of technology (1st ed.). Oxon, United Kingdom: Routledge.

Amabile, T.M. & Conti, R. & Coon, H. & Lazenby, J. & Herron, M. (1996). Assessing the Work Environment for Creativity. Academy of Management, 39(5), 1154–1184.

Becker, R., & Patnaik, D. (1999). Needfinding: The Why and How of Uncovering People's Needs. Design Management Journal, 10(2), 37–43.

Beckman, S. L. & Barry, M. (2007). Innovation as a Learning Process: Embedding Design Thinking. California Management Review, 50(1), 25–56

Beckman, S. C. & Langer, R. (2005). Sensitive research topics: netnography revisited. Qualitative Market Research: An International Journal, 8(2), 189–203.

de Bono, E. (1999). Six thinking hats (2nd ed.). Boston, MA: Back Bay Books.

Brenda, L. (Ed.). (2003). Design Research: Methods and Perspectives. Cambridge, MA: MIT Press.

Brill, M., Weidemann, S., & BOSTI Associates (2001). Disproving widespread myths about workplace design. URL: https:/www.researchgate.net/publication/248708306_DISPROVING_WIDESPREAD_MYTHS_ABOUT_WORKPLACE_DESIGN?ev=srch_pub (20.03.2015)

Broderick, A. J. & Lee, N. (2007). The past, present and future of observational research in marketing. Qualitative Market Research: An International Journal, 10(2), 121-9.

Brown, T. (2009). Change by design: How design thinking transforms organizations and inspires innovation. HarperBusiness.

Brown, S., Gray, D., & Macanufo, J. (2010). Gamestorming: A playbook for innovators, rulebreakers, and changemakers. Sebastopol, CA: O'Reilly.

Buchanan, R. (1992). Wicked Problems in Design Thinking. The MIT Press, 8(2), 5-21.

Buxton, B. (2007). Sketching User Experiences — getting the design right and the right design. Amsterdam: Elsevier Inc.

Capodagli, B. (2010). Magic in the Workplace: How Pixar and Disney Unleash the Creative Talent of Their Workforce. Effectif, 13(4).

Clark, T., Osterwalder, A., & Pigneur, Y. (2010). Business model generation: A handbook for visionaries, game changers, and challengers (1st ed.). Hoboken, NJ: John Wiley and Sons.

CCleaner. (n.d.). PC Optimization and Cleaning. Retrieved February 11, 2015, from https://www.piriform.com/ccleaner

Cooper, A., & Reimann, R. (2003). About face 2.0: The essentials of interaction design (2nd ed.). New York, NY: Wiley.

Creuznacher, I., & Grots, A. (2011). Design Thinking. Querdenker, (3), 74–78.

Deutsche Bank Annual Report 2013 — Strategy 2015+. URL: https://annualreport.deutsche-bank.com/2013/ar/deutsche-bank-group/corporate-profile/strategy-2015.html (15.03.2015)

Doorley, S., & Witthoft, S. (2012). Make space: How to set the stage for creative collaboration. Hoboken, NJ: John Wiley & Sons.

D.School Stanford University (2015). What to do in Need Finding. URL: http://hci.stanford.edu/courses/dsummer/handouts/NeedFinding.pdf (01.02.2015)

D.School Stanford University (2015b). bootcamp bootleg. URL: dschool.stanford.edu/wp-content/uploads/2011/03/BootcampBootleg2010v2SLIM.pdf (01.02.2015)

Emerson, R. M., Fretz, R. I., & Shaw, L. L. (1995). Writing ethnographic fieldnotes. Chicago, IL: University Of Chicago Press.

Eppler, M. J., Kernbach, S., Wiederkehr, B., Gassner, P. (2014). The Confluence Diagram: Embedding Knowledge in Interaction Constraints. Proceedings of the IEEE Information Visualization Conference (InfoVis), 9–14 November 2014, Paris, France.

ESOMAR (2007). ICC/ESOMAR International Code on Market and Social Research. https://www.esomar.org/uploads/public/knowledge-and-standards/codes-and-guidelines/ICCESOMAR_Code_English_.pdf (18.6.2015)

FabLab Zürich. (2012). URL: http://zurich.fablab.ch/fab-charta

Florida, R. (2014). The Rise of the Creative Class — Revisited: Revised and Expanded. New York, NY: Basic Books.

Gaegauf, M. G. (2014). Office Design for Introverts & Extroverts: A concept to create integrated work spaces for introverts and extroverts to enhance productivity and creativity (Unpublished thesis). Hochschule Luzern: Design & Kunst, Luzern, Switzerland.

Groves, K. (2013). Four types of space that support creativity & innovation in business. URL: http://enviableworkplace.com/four-types-of-space-that-support-creativity-innovation-in-business/ (18.02.2015)

Guest, G., Mack, N., MacQueen, K., Maney, E. & Woodsong, C. (2005). Qualitative Research Methods: A Data Collector's Field Guide. Research Triangle Park, NC: Family Health International.

Hahn, H. (2015). Hilary Hahn acceptance speech [Television series episode]. Interview. In The 57th Grammy's. Los Angeles, CA.

Hauschildt, J. (2004). Innovationsmanagement. München: Verlag Franz Vahlen, S.317

Herrmann International. (n.d.). URL: http://www.herrmannsolutions.com (19.03.2015)

Herstatt, C., & von Hippel, E. (1992). From experience: Developing new product concepts via the lead user method: A case study in a "low-tech" field. Journal of Product Innovation Management, 9, 213–221.

Herstatt, C., and Lüthje, C. (2004). The Lead User method: an outline of empirical findings and issues for future research. R&D Management, 34: 553–568.

International Chamber of Commerce & ESOMAR. (2008). ICC/Esomar International Code on Market and Social Research. URL: http://www.vsms-asms.ch/files/7113/5625/5208/ICCESOMAR_Code_German_2008.pdf (20.03.2015)

Kahler Slater (2010). What makes a great workplace? Learning from the best place to work companies. URL: http://www.kahlerslater.com/content/pdf/What-Makes-a-Great-Workplace-white-paper.pdf (20.03.2015)

Kelley, T., & Littman, J. (2001). The Art of Innovation: Lessons in Creativity from IDEO (1st ed.). New York, NY: Doubleday Religious Publishing Group.

Kerr, S., & Landauer, S. (2008). Using stretch goals to promote organizational effectiveness and personal growth: General Electric and Goldman Sachs. The Academy of Management Executive, 18(04), 134-138.

Kerstan, P., (Regie) & Schmuckler, A. (Regie). 1972. Richtung 2000 - Vorschau auf die Welt von morgen (Television broadcast), Germany: ZDF.

Kim, W. C., & Mauborgne, R. (2005). Der blaue Ozean als Strategie: Wie man neue Märkte schafft, wo es keine Konkurrenz gibt. München: Hanser.

Knoll Workplace Research (2012). Activity Spaces: A variety of spaces for a variety of work. Retrieved from: http://www.knoll.com/media/870/279/wp_ActivitySpaces.pdf

Knoll Workplace Research (2013). Creating Collaborative Spaces that Work. Retrieved from: https://www.knoll.com/knollnewsdetail/creating-collaborative-spaces-that-work

Kubrick, S. (Regie). (1968). 2001: A Space Odyssey [Motion picture]. United States: Metro-Goldwyn-Mayer.

Kuniavsky, M. (2003). Observing the user experience: A practitioner's guide to user research. San Francisco, CA: Morgan Kaufmann Publishers.

Krueger, R. A., & Casey, M. A. (2000). Focus groups: A practical guide for applied research (p. 87). Thousand Oaks, CA: Sage Publications.

Leifer, L., Steinert, M. (2011). Dancing with ambiguity: Causality behaviour, design thinking, and triple-loop-learning. Information, Knowledge, Systems Management. Volume, 10, 1-4.

Martin, R. (2013). Rethinking the Decision Factory — HBR. URL: https://hbr.org/2013/10/rethinking-the-decision-factory (22.03.2015)

McGregor, D. (2006). The Human Side of Enterprise (annotated ed.). New York: McGraw-Hill.

Meinel, C., Plattner, H., & Weinberg, U. (2009). Design Thinking. München, Deutschland: FinanzBuch Verlag GmbH.

Meinel, C., Weinberg, U., & Krohn, T. (2015). Design Thinking Live: Wie man Ideen entwickelt und Probleme löst. Hamburg: Murmann Publishers.

Meel, J., Martens, Y., & Ree, H. J. (2010). Planning office spaces: A practical guide for managers and designers. London, U.K: Laurence King.

M Lab. (n.d.). Ethics in Design - A Quick Primer. Retrieved February 11, 2015, from http://mlab.uiah.fi/polut/Yhteiskunnalliset/lisatieto_ethics_primer.html

Morgan, G. (1997). Images of organization (2nd ed.). Thousand Oaks, CA: Sage Publications.

Morgan, G. (1993). Imaginization: The art of creative management. Newbury Park, CA: Sage Publications.

Office Snapshots - i know where you work. (n.d.). URL: http://officesnapshots.com (12.02.2015)

Office Snapshots - i know where you work. [Photographs]. URL: http://officesnapshots.com (20.03.2015)

Ohno, T. (1988). Toyota production system: Beyond large-scale production. Cambridge, MA: Productivity Press.

O'Neill, M., & Wymer, T. (2010). Implementing Integrated Work to Create a Dynamic Workplace. URL: http://www.knoll.com/media/909/960/WP_ImplementingIntegratedWork.pdf (20.03.2015)

Parker, G. M. (2008). Team Players and Teamwork: New Strategies for Developing Successful Collaboration. (2nd ed.). San Francisco, CA: John Wiley & Sons.

Pautsch, P., & Steininger, S. (2014). Lean Project Management: Projekte exzellent umsetzen. München: Hanser.

PechaKucha 20x20. (n.d.). URL: from http://www.pechakucha.org (01.02.2015)

Pine, B. J., & Gilmore, J. H. (1999). The experience economy: Work is theatre & every business a stage. Boston: Harvard Business School Press.

Pinterest. (n.d.). URL: http://www.pinterest.com (20.03.2015)

Reckwitz, A. (2012). Die Erfindung der Kreativität: Zum Prozess gesellschaftlicher Ästhetisierung. Berlin: Suhrkamp Verlag.

Rittel, H.W. & Webber, M.M. (1973). Dilemmas in a General Theory of Planning. Policy Sciences, 4, 155–169.

Rogers, E.M. (2003). Diffusion of Innovations (5th ed.). New York, NY: Free Press.

Sarasvathy, S. D. (2008). Effectuation: Elements of entrepreneurial expertise. Cheltenham, UK: Edward Elgar.

Schar, M. (2010). Team Formation. Presentation at Stanford University.

Schindlholzer, B. (2014). Methode zur Entwicklung von Innovationen durch Design Thinking Coaching. St. Gallen, Switzerland: D-Druck Spescha.

Shapeways - 3D Printing Service and Marketplace. (n.d.). Retrieved from http://www.shapeways.com

Silver, C. (2011). Needfinding 101 [PowerPoint]. URL: http://wikibox.stanford.edu/14-15/index.php/Resources Lectures?action=download&upname=5NovCaraSilver_persona_excerpts.pdf (20.03.2015)

Stanford University. (n.d.). d.school Software Design Experiences.

Steelcase. (2014). Insights Applied. 360 Magazine Future Focused, (64), 112. URL: http://360.steelcase.com/issues/future-focused-2/ (20.03.2015)

Steelcase. (n.d.). Susan Cain Quiet Spaces by Steelcase. URL: http://www.steelcase.com/en/products/category/architectural/archwalls/via/pages/quiet-spaces.aspx (17.02.2015)

Stegmeier, D. (2008). Innovations in Office Design: The Critical Influence Approach to Effective Work Environments. Hoboken: John Wiley & Sons.

Steve Jobs. (2013). ‚Innovation Distinguishes Between A Leader And A Follower‘ - Forbes. URL: http://www.forbes.com/sites/bwoo/

2013/02/14/innovation-distinguishes-between-a-leader-and-a-follower/ (20.03.2015)

Surowiecki, J. (2004). The Wisdom of Crowds. New York, NY: Anchor.

Uebernickel, F., & Brenner, W. (2015). Business Innovation: Das St. Galler Modell. Springer Gabler. (Erscheinungsdatum 5. August 2015)

Ulrich, K. T., & Eppinger, S. D. (1995). Product design and development (4th ed.). New York, NY: McGraw-Hill.

Verein Deutscher Ingenieure. (2004). Methodisches Entwerfen technischer Produkte VDI 2223. Berlin, Deutschland: Beuth.

Vetterli, C., Brenner, W., Uebernickel, F., & Petrie, C.(2013). From Palaces to Yurts - Why Requirements Engineering Needs Design Thinking. In: IEEE INTERNET COMPUTING 17 (2013), Nr. 2, S. 91–94.

Vetterli, C., Uebernickel, F., Brenner, W., Haeger, F., Kowark, T., Krueger, J., Mueller, J., Plattner, H., Stortz, B., & Sikkha, V. (2013b). Jumpstarting Scrum with Design Thinking. St.Gallen: University of St.Gallen.

Walker, T. (2013, September 20). Perks for employees and how Google changed the way we work (while waiting inline). URL: http://www.independent.co.uk/news/world/americas/perks-for-employees-and-how-google-changed-the-way-we-work-while-waiting-in-line-8830243.html (12.02.2015)

Wilde, D.J. (2007). Teamology: The Construction and Organization of Effective Teams. Stanford University.

Wymer, T. (2010). Proportional Planning for the Adaptable Workplace. URL: http://www.knoll.com/media/309/821/WP_ProportionalPlanning_AdaptableWorkplace.pdf (20.03.2015)

Zydra, M. (2014). Design Thinking - Labor für Geistesblitze. URL: http://www.sueddeutsche.de/wirtschaft/Design Thinking-in-unternehmen-labor-fuer-geistesblitze-1.1856849 (20.03.2015)

3D Hubs. (n.d.). URL: www.3dhubs.com (20.03.2015)

3D Printing Service i.materialise. (n.d.). URL: http://i.materialise.com (20.03.2015)

INDEX

THE AUTHORS

Falk Uebernickel
Prof. Dr. oec. Falk Uebernickel is a professor at the Hasso-Plattner-Institute (HPI) in Potsdam (Germany) and an adjunct professor at the University of St.Gallen (HSG) in St. Gallen (Switzerland). He is teaching Design Thinking at the master student level together with Stanford University and is conducting executive lectures at companies. Alongside, he has worked for companies globally in various industries such as banking, insurance, automotive, sports, pharmaceutical and many more on over 80 different projects. Furthermore, Prof. Uebernickel is the spokesperson of the SUGAR network — a global movement and initiative of over 20 universities and more than 100 companies to apply design thinking to real-world challenges. His focus is on training and teaching as well as on strategic planning of digital transformation projects in companies.

Li Jiang
Dr. Li Jiang is the Director of Stanford AIRE (AI, Robotics and Education), and has been doing research in the field of Robotics and AI for many years. He won the top award Best of Innovations at the 2014 Consumer Electronics Show (CES) Innovation Awards as well as the top award Best of the Program at the National Lincoln Design Competition. Dr. Jiang has served on the Judging Committee for the CES Innovation Awards, and he is a board member of the International Robotic Expert Committee for China (IRECC). His current research focuses on the ways that AI and Robotic Technologies will impact our education system and how we need to adjust the existing system to accommodate the coming era of Robotics and AI. His class AI, Robotics and Design of Future Education is the first class at Stanford University to address this multidisciplinary field of AI, Robotics, Education and Design Innovation. Dr. Jiang attended Stanford University where he earned a Master's Degree in Design Innovation, a Ph.D. in Robotics, and a Ph.D. minor in Management Science. He holds more than 50 U.S. and international patents.

Walter Brenner
Since April 1st, 2001, Prof. Dr. oec. Walter Brenner, is Professor of Information Management at the University of St. Gallen and Managing Director of the Institute of Information Management. Before that, he held professorships at the University of Essen and the TU Bergakademie Freiberg. His research fields are: Industrialization of Information Management, IT Service Provider Management, Customer Relationship Management, New Technologies and design thinking. Next to this he is also a freelance consultant on issues of information management and business preparation for the digital, connected world.

Britta Pukall
Britta Pukall, owner of the leading industrial design agency milani design & consulting AG in Zurich-Thalwil and Berlin. On the one hand, the focus of her work is on innovation development for market leaders in a wide variety of industries — using design thinking alongside other approaches. On the other, she supports start-ups and more mature companies in terms of their strategic and emotional positioning, because only when this is known one can generate suitable innovations. With her interdisciplinary team, she ensures that the innovation approaches and strategies are implemented in detail. Britta Pukall studied Design and Architecture (Magister Artium) in Kassel and Vienna and completed an Executive MBA at the University of St. Gallen.

Therese Naef
Therese Naef initially completed a technical model-making apprenticeship and then led an architectural model construction branch. Subsequently she studied product design in Zurich (Dipl. Des.). For more than 10 years, she is the Managing Director and Managing Partner of the leading Swiss industrial design agency milani design & consulting AG in Zurich-Thalwil. Above all, she is responsible for the implementation of customer and in-house projects. With her understanding of both, content-related and implementation-oriented processes and solutions, she forms the necessary bridge for companies between strategy, innovation and tangible realizability.

Bernhard Schindlholzer
Dr. oec. Bernhard Schindlholzer is Product Manager at Google. Prior to that, he founded several companies in the areas of Digital Product Development and User Experience Design.
In his dissertation at the University of St. Gallen, he researched the practical application of design thinking and documented it in the context of a comprehensive methodology.

DESIGN THINKING TOOLBOX

The toolbox is a collection of methods for systematically solving problems during design thinking. As a guideline, it offers beginners support in identifying the right method in each particular phase.

Experienced design thinkers can use the toolbox as a source of inspiration.

This table lists all the methods found in this handbook and orders them within the steps of the micro-cycle and the process phases of the macrocycle as described in the book. The indicators show how appropriate the authors regard each method as being for each phase. Where there is no indication, methods may be conditionally appropriate to phases. Finding each method's description in the book is supported by page number listing.

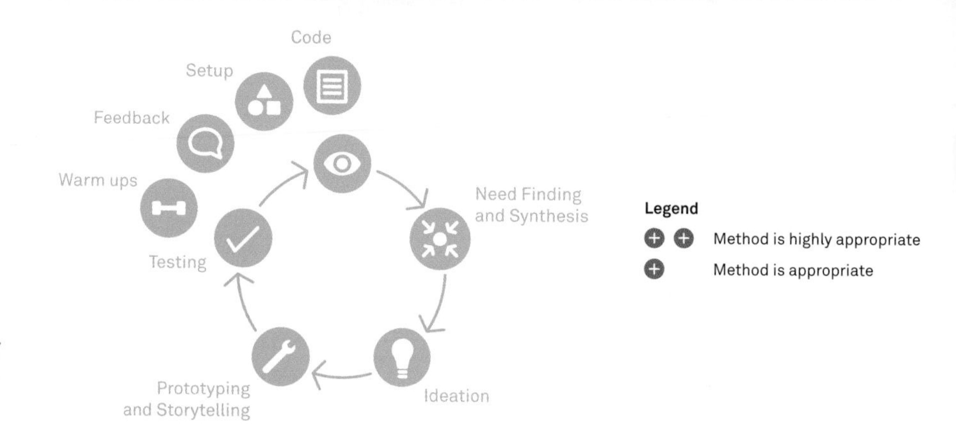

Category	Technique	Page	Design Space Exploration	Critical Function	Darkhorse Prototype	Funky Prototype	Functional Prototype	X-is-Finished	Final Prototype
Setup	Project planning	64	++			+	+		
	Infrastructure	71	++						
	Team setup	73	++						
	Post-its	78	++	++	++	++	+	+	+
	DNA analysis method	82							
Problem definition and re-definition	Definition of problem statement	88	++						
	Stretch goals	90			++	+			
	Framing and re-framing	92	+	+	++	+			
	Get inspriration from the future	94			++				
Need finding and synthesis	Need finding cycle	98	++	++	++	++	+	+	+
	Guidelines for formulation	100	++						
	Sampling techniques	101	++	++	++	+	+	+	+
	Target group identification	102	++	++	++	+	+	+	+
	Focus groups	104	+	++	++	+	+	++	++
	Interviews	106	++	++	++	++	+	+	+
	Observation	109	++	++	++	+			
	Engagement	111	++	++	++	+			
	Benchmarking	112	++	++	+				
	Frameworks	114	+	+	+	++			
	Field notes	116	++	++	++	++	++	++	
	Moodboard	118	++	+	+	+			
	AEIOU	120	++	++	++	+			
	Empathy map	122	++	++	++	+			
	Netnography	124	+	++	++	+			

	Persona	125	⊕ ⊕	⊕ ⊕	⊕				
	Why-how laddering	128	⊕ ⊕	⊕ ⊕	⊕ ⊕	⊕ ⊕	⊕ ⊕		
	5 Whys	129	⊕	⊕ ⊕	⊕ ⊕	⊕ ⊕			
	Point of view	130	⊕ ⊕	⊕ ⊕	⊕ ⊕	⊕			
	Lead user	132	⊕ ⊕	⊕ ⊕	⊕ ⊕				
	Camera study	133	⊕	⊕ ⊕	⊕ ⊕				
Ideation	Brainstorming	138	⊕	⊕ ⊕	⊕ ⊕	⊕ ⊕	⊕	⊕	
	Brainwriting	140	⊕	⊕ ⊕	⊕ ⊕	⊕ ⊕	⊕	⊕	
	Six thinking hats	142	⊕	⊕ ⊕	⊕ ⊕	⊕ ⊕			
	Power of ten	144	⊕	⊕ ⊕	⊕ ⊕	⊕ ⊕	⊕	⊕	
	How might we	145	⊕	⊕ ⊕	⊕ ⊕	⊕ ⊕	⊕	⊕	⊕
Prototyping and storytelling	Wireframing	152	⊕	⊕ ⊕	⊕ ⊕	⊕			
	Mock-ups	154				⊕	⊕ ⊕	⊕ ⊕	⊕
	Open hardware	156		⊕		⊕	⊕ ⊕	⊕ ⊕	⊕ ⊕
	Roleplays	158	⊕	⊕ ⊕	⊕ ⊕	⊕ ⊕	⊕ ⊕	⊕ ⊕	⊕ ⊕
	Bodystorming	160	⊕	⊕ ⊕	⊕ ⊕	⊕ ⊕	⊕ ⊕	⊕ ⊕	⊕ ⊕
	Paper prototyping	162	⊕ ⊕	⊕ ⊕	⊕ ⊕	⊕			
	Storytelling and storywriting	164	⊕ ⊕	⊕ ⊕	⊕ ⊕	⊕ ⊕	⊕ ⊕	⊕ ⊕	⊕ ⊕
	Comics	166	⊕	⊕ ⊕	⊕ ⊕	⊕			
	3D rapid prototyping	168				⊕	⊕ ⊕	⊕ ⊕	⊕ ⊕
	Video prototyping	170		⊕	⊕	⊕	⊕ ⊕	⊕ ⊕	⊕ ⊕
	Service blueprinting	172		⊕ ⊕	⊕ ⊕	⊕	⊕		
	Sketches and scribbles	174	⊕ ⊕	⊕ ⊕	⊕ ⊕				
	Photoshop prototypes	176		⊕ ⊕	⊕ ⊕	⊕			
	Combined prototypes	178		⊕	⊕	⊕ ⊕	⊕ ⊕		
	Town planning	180		⊕ ⊕	⊕ ⊕	⊕ ⊕	⊕ ⊕	⊕ ⊕	⊕ ⊕
	Business model prototypes	180		⊕	⊕	⊕	⊕ ⊕	⊕ ⊕	⊕ ⊕
	Confluence dynagram	182					⊕	⊕ ⊕	⊕ ⊕
Testing	Consumer clinics	186					⊕	⊕	⊕ ⊕
	Usability testing	187					⊕	⊕ ⊕	⊕ ⊕
	NABC Pitch	188	⊕	⊕	⊕	⊕	⊕	⊕	⊕
	PechaKucha	189	⊕	⊕	⊕	⊕	⊕	⊕	⊕
Warm ups	Spaghetti tower	194	⊕ ⊕	⊕ ⊕	⊕ ⊕	⊕ ⊕	⊕ ⊕	⊕ ⊕	⊕ ⊕
	Yes but, yes and	196	⊕ ⊕	⊕ ⊕	⊕ ⊕	⊕ ⊕	⊕ ⊕	⊕ ⊕	⊕ ⊕
	Races	198	⊕ ⊕	⊕ ⊕	⊕ ⊕	⊕ ⊕	⊕ ⊕	⊕ ⊕	⊕ ⊕
	Assembly	200	⊕ ⊕	⊕ ⊕	⊕ ⊕	⊕ ⊕	⊕ ⊕	⊕ ⊕	⊕ ⊕
	Stick figures	202	⊕ ⊕	⊕ ⊕	⊕ ⊕	⊕ ⊕	⊕ ⊕	⊕ ⊕	⊕ ⊕
Feedback	I like, I wish, what if	206	⊕ ⊕	⊕ ⊕	⊕ ⊕	⊕ ⊕	⊕ ⊕	⊕ ⊕	⊕ ⊕
	Plus or delta	208	⊕ ⊕	⊕ ⊕	⊕ ⊕	⊕ ⊕	⊕ ⊕	⊕ ⊕	⊕ ⊕
	Feedback capture grid	210	⊕ ⊕	⊕ ⊕	⊕ ⊕	⊕ ⊕	⊕ ⊕	⊕ ⊕	⊕ ⊕
	Critical reading checklist	212	⊕ ⊕	⊕ ⊕	⊕ ⊕	⊕ ⊕	⊕ ⊕	⊕ ⊕	⊕ ⊕
Design thinking code	Declaration of consent	213	⊕ ⊕	⊕ ⊕	⊕ ⊕	⊕ ⊕	⊕ ⊕	⊕ ⊕	⊕ ⊕